Advances in Wastewater Treatment I

Edited by

Vimal Gandhi[1] and Kinjal Shah[2]

[1]Department of Chemical Engineering, Faculty of Technology, Dharmsinh Desai University, Nadiad-387 001, Gujarat, India

[2]College of Urban Construction, Department of Municipal Engineering, Nanjing Tech University (NTU), Nanjing, China 218 816

Published by **Materials Research Forum LLC**
Millersville, PA 17551, USA

Published as part of the book series
Materials Research Foundations
Volume 91 (2021)
ISSN 2471-8890 (Print)
ISSN 2471-8904 (Online)

Print ISBN 978-1-64490-114-4
eBook ISBN 978-1-64490-115-1

Distributed worldwide by

Materials Research Forum LLC
105 Springdale Lane
Millersville, PA 17551
USA
https://www.mrforum.com

Manufactured in the United States of America
10 9 8 7 6 5 4 3 2 1

Table of Contents

Preface

United Nations General Assembly set 17 different goals for the sustainable development (SD) of the world with the time frame of 2030. Out of 17 SD goals, 8 goals are directly or indirectly related to water in terms of its availability, municipal and industrial waste water treatment, its reuse and recycle policy. Out of total global water consumption, approximately 25% is attributed to industrial consumption. It is nearly 40-45% in developed countries as compared to 15-20% consumption rate in developing countries. As per one estimate, globally 30-50% of the industrial wastewater flows back into the ecosystem either being poorly treated or without being treated or reused depending upon the types of industries. This discharged water is adversely affecting the ecosystem and the environment. This statistics indicates the importance of water and need of its treatment technologies for the coming decade.

At present conventional waste water treatment including primary and secondary treatment methods are widely being used for the treatment of industrial or municipal waste water. Due to rapid growth of industrialization and urbanization, the quality of waste water is also changing significantly and the world is facing problems of removal of various complex organic compounds from the waste water. Conventional treatment technologies are not efficient to treat this waste water effectively and meet the discharge norms laid down by different regulatory agencies worldwide. It is the need of the present era to find out alternative treatment methods in the collaboration of conventional methods to treat the industrial and municipal waste water effectively. Worldwide researchers are focusing on these burning issue of how to develop various technologies which resulted into improvement in conventional treatment technologies. In recent past, there are various alternative materials and methods proposed in the literature as part of advances in waste water treatment technologies. These includes mainly (i) Advanced Oxidation Processes (ii) Various types of Membrane Technologies (iii) Detection and removal of heavy metals and organic compounds (iv) Explore the possibility of nanomaterials in waste water treatment (v) Low cost adsorbents and bio flocculants (vi) Policy making framework for waste water treatment. The basic objective of the proposed book "Advances in Waste Water Treatment- Part I" is to address the advancement in waste water treatment technologies which try to include majority of the above mentioned points in the different chapters.

Chapter 1 summarizes conventional waste water treatment technologies bringing out the need and highlighting the types of advance waste water treatment technologies. Advanced Oxidation Processes (AOPs) are gaining importance due to their ability to degrade complex organic molecules into less harmful compounds like carbon dioxide and water. **Chapter 2** focusing on review of various advanced oxidation processes with suitable mechanisms involved. **Chapter 3** emphasizes on application of one of the AOPs, photocatalytic degradation of pharmaceutical compounds from waste water using nanocrystalline titanium dioxide.

Chapter 4 comprises of overview of removal of arsenic from water with various treatment methods including membrane technology. Quantitative arsenic analysis includes several colorimetric, luminescence, spectroscopic, atomic absorption, mass spectrometric and biosensor based techniques have been highlighted in the same chapter.

Adsorption process is also widely used as a part of tertiary treatment or polishing methods to treat waste water. Apart from conventional activated charcoal, research community mainly focused on developing alternative low cost adsorbent and how to develop immobilized microorganism on conventional powdered activated carbon to improve its effectiveness for the removal of organic molecules from waste water. **Chapter 5** is focusing on applicability of low cost adsorbent derived from egg shells in waste water treatment. **Chapter 6** highlights new concept of immobilized microorganism on powdered activated carbon for the degradation of polycyclic aromatic hydrocarbons (PAHs) namely pyrene. **Chapter 7** covers various types of bio flocculants used for the waste water treatment, which is one of the alternative flocculants to conventional chemicals used in primary treatment.

This volume is indeed the results of remarkable cooperation of many distinguished experts for contributing their review or research article. We are very much thankful to all the authors for their willingness to provide book chapters and join hands with us to create awareness about advances in waste water treatment. We would like to express our gratitude to all the publishers and authors for granting us the copyright permission to reprint the same. Although sincere efforts have been made to get copyright permissions from the respective agencies, we would sincerely apologies to any copyright holder if unknowingly their right is being infringed.

We would like to take this opportunity to express our sincere gratitude towards Dr. H.M.Desai Sir(Vice Chancellor, Dharmsinh Desai University-DDU), Dr. P. A. Joshi sir and Dr. M.S.Rao sir for their continuous motivations and guidance. We would acknowledge the sincere efforts of Mr. Thomas Wohlbier for his support from proposal stage to final form of edited book. Our sincere thanks to Dr. P.A.Joshi, Dr. Krishnakumar and Dr. Parasuraman for their supports in review the book chapters and provided valuable suggestions. We are also thankful to staff members of Department of Chemical Engineering (Dharmsinh Desai University, Nadiad) and group members of KJS Lab (Nanjing Tech University, China) for their support in the preparation of book. We are also thankful to all the friends and family members for their best wishes.

Dr. Kinjal Shah
College of Urban Construction,
Department of Municipal Engineering
Nanjing Tech University (NTU),
Nanjing, China 218 816

Dr. Vimal Gandhi
Department of Chemical Engineering,
Faculty of Technology,
Dharmsinh Desai University,
Nadiad-387 001, Gujarat, INDIA

Advances in Wastewater Treatment I
Materials Research Foundations **91** (2021) 1-36

Materials Research Forum LLC
https://doi.org/10.21741/9781644901144-1

Chapter 1

Introduction to Conventional Wastewater Treatment Technologies: Limitations and Recent Advances

Gangaraju Gedda[1*], Kolli Balakrishna[2], Randhi Uma Devi[3], Kinjal J Shah[4]

[1] Department of Basic Science, Vishnu Institute of Technology, Vishnupur, Bhimavaram-534202, A.P, India

[2] Department of Chemistry, GITAM (Deemed to be University), Visakhapatnam, India-530045

[3] Department of Chemistry, School of Science, GITAM (Deemed to be University), Hyderabad, Telangana-502329, India

[4] College of Urban Construction, Nanjing Tech University, Nanjing, 211800, PR China

*raju.analy@gmail.com

Abstract

The rapid growth of the industries and population leads to increasing generation of industrial and municipal wastewater. This wastewater threatens directly or indirectly the human health and industrial processes. Therefore, it is necessary to develop a rapid, simple, eco-friendly, effective, and efficient method for eliminating pollutants from industrial and municipal wastewater. The wastewater treatment aims to remove pollutants including particles, organic/inorganic substances, and pathogenic microorganisms, and finally returned to the cycle. This chapter presents a brief introduction to the issue associated with municipal and industrial wastewater. Also, this chapter presents detailed information about the conventional wastewater treatment methods. Specifically, it discusses the steps involved in the wastewater treatment *viz.* primary, secondary, and tertiary treatment.

Keywords

Wastewater, Pollutant, Treatment, Industrial, Municipal

Contents

1. Introduction

Water is an essential commodity for the survival of all living species [1]. Especially, human beings are using the available groundwater as well as surface water for various purposes including drinking, irrigation, washing, cooking, and different industrial operations [2]. However, water should fulfill the minimum criteria of quality as per the world health organization (WHO) standards [3]. But the water is not fulfilling standards due to changes in its characteristics (physical, chemical and even in biological). This circumstance makes water not a suitable candidates for drinking as well as other purposes. This alteration of water properties is termed as water pollution or wastewater. Water pollution is mainly due to either industrial waste water or municipal wastewater [4]. Different industrial as well as municipal wastewater sources are shown in Fig. 1.

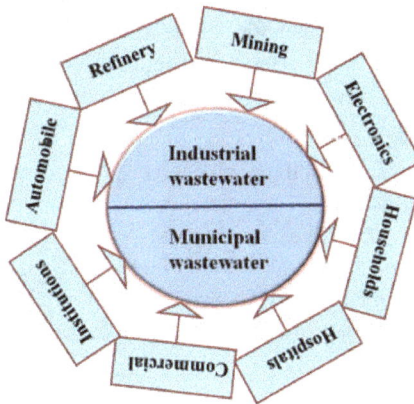

Figure 1. Schematic illustration of industrial and municipal wastewater sources.

Generally, municipal wastewater is generated from households, public facilities (restaurants & malls), and agricultural activities (fertilizers, insecticides, fungicides, and herbicides) [5]. Meanwhile, industrial wastewater is generated from fresh water which is used as integrated part of chemical process industries for various activities starting from raw material stage to product finishing stage. As an outcome, wastewater of different characteristics are generated depending upon the types of industries like petroleum refinery, petrochemical, fertilizers, fine chemicals, pharmaceuticals, etc. [6]. Industrial wastewater is adversely affecting human health and its environment as it content toxic heavy metals, suspended solids, complex organic compounds [7-11]. Hence, it is essential to develop a rapid, simple, effective, and efficient method for the treatment of industrial and municipal wastewater.

The proposed chapter is to give emphasis to conventional wastewater treatment methods like preliminary, primary and secondary treatment. Apart from these, various tertiary treatment methods are also discussed including membrane-based methods, electrodialysis, adsorption method, etc.

2. The water treatment process for the municipal and industrial sector

The aim of the wastewater treatment method is to decrease/remove inorganic and organic components, toxic substances, killing pathogenic microorganisms, etc. [12-18]. Consequently, the quality of treated water is enhanced to meet the requirements of WHO standards or pollution control authority of the respective country. The pollutants of both municipal and industrial wastewater differ from place to place [19-21]. Therefore, the types of treatment of wastewater rely on its nature and the required quality of water after the treatment.

The wastewater treatments usually involve three stages, namely primary, secondary, and tertiary treatments (Fig. 2) [4]. The primary and secondary treatment process is used to removing the majority of large particles and organic matter, respectively. After the primary and secondary treatments, various undesirable matters remain in the treated water, the tertiary treatment used to remove such matter as a polishing unit. These treatments generally involve a combination of different physical, chemical, and biological processes.

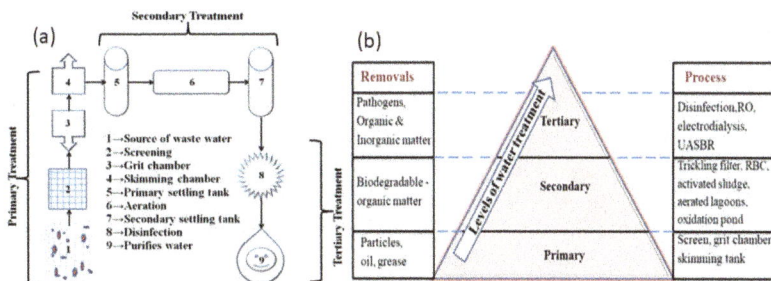

Figure 2 (a) Flow diagram of primary, secondary and tertiary treatment process; (b) Flow chart for levels of wastewater treatment

2.1   Primary treatment process

This process includes two consecutive steps *viz.* preliminary treatment as well as sedimentation process [22-24]. Preliminary treatment mainly consists of screening, grit chamber and skimming tank for the removal of large particles and debris from wastewater, oil and fats. Further, the wastewater is subjected to sedimentation or chemical precipitation in primary settling tanks, which removes organic solids, colloidal and finer suspended particles in the form of sludge.

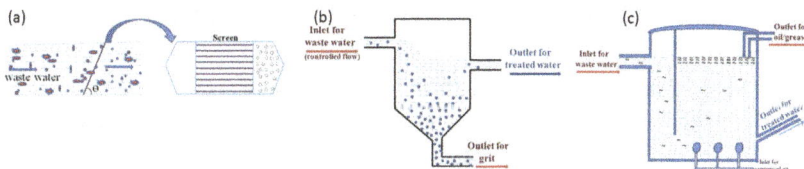

Figure 3. Primary treatment process (a) Screening process (b) Grit Chamber and (c) Skimming tank of wastewater.

2.1.1 Screen

The first step in the wastewater treatment is to remove entrained large, floating, or suspended solids, these may consist of pieces of wood, plastic, cloth, and paper. The main objective of screening is to protect other mechanical equipment and to prevent clogging of valves and other accessories in the wastewater treatment plant [25-28].

As shown in Fig. 3a, screens are in rectangular with the opening of uniform in size with a perforated metal plate. Screens are arranged at an inclined angle (θ) of $30° - 60°$ to the flow direction of the wastewater [29-31]. There are various types of screens available based on the size of the opening *viz.* coarse, medium, and fine screen.

2.1.2 Grit chambers

Grit chamber is used to eliminate grit such as sand, metal fragments and broken glass pieces etc. from wastewater [32, 33]. It is important to remove such kind of materials from wastewater to protect pumps and such kind of mechanical equipment from wear and tear. The grit chamber is a long narrow rectangular channel or long basin works as sedimentation tank to take care of the removal of heavier inorganic materials and to pass forward lighter organic materials for the next processing steps (Fig. 3b). It is designed to control the velocity of the flow of wastewater to remove the material [33-35]. If the flow rate is too low or high, it causes the settling of lighter organic matter or settling of the entire grit particles, respectively. Therefore, a velocity control device "proportional flow weir" is arranged at the outlet of the grit chamber to maintain constant flow velocity. Various types of grit removal chambers including aerated grit chambers, vortex-type grit removal system, detritus tanks, horizontal flow grit chambers are utilized in wastewater treatment plants depending upon characteristics of wastewater.

2.1.3 Skimming tanks

A skimming tank is used to remove floating matter such as oil, fat, wax, soaps, and grease [36]. This floating matter adversely affecting the performance of the activated sludge process and inhibits the biological growth in further treatment stages. The skimming tank is a long trough-shaped chamber divided into two or three lateral compartments by vertical baffle walls (Fig. 3c)[35]. The wastewater enters the tank from the inlet, flows longitudinally, and leaves by a narrow inclined channel. Air blowers are arranged at the bottom of the tank to raise the floating matter which remains on the surface of the water, which can then be easily removed.

2.1.4 Primary Settling Tanks (PST)

Primary settling tanks are used for the removal of remaining organic matter and suspended particles from wastewater [13]. The presences of these substances are leads to more oxygen demand to the receiving waterbody, interferes with the biological treatment, and reduces disinfection efficiency in the fallowing treatment process [37]. This step involves the physical and chemical sedimentation processes for the affective elimination of fine particles.

The physical sedimentation is the strategy of removing heavier suspended particles by allowing the wastewater undisturbed in the tank for ~2-6 hours. During this time, the heavier particles settled down at the bottom of the tank due to the gravitational force. The settled solid partials are termed as sludge. The clear supernatant water (effluent) is then drawn from the tank with the help of pumps and used for further chemical treatment.

Even after the physical sedimentation process, the effluent still contains finer and colloidal particles. Hence chemical sedimentation process used to remove unsettled matter by using coagulants like alum, ferric chloride, ferric sulfate, etc. The addition of a coagulant leads to the formation of floc, which will trap the fine particles and form agglomerate, convert it into heavier and settled down at the bottom of the tank. This treatment method with proper coagulant dosages may be expected to remove 45-65 % of biological oxygen demand (BOD) and 60-80 % of suspended solids [38-40].

2.2   Secondary treatment

In the secondary treatment process, biodegradable soluble organics compounds are degraded through microorganisms [41-42]. Suitable types of microogranisms feed to the wastewater, increasing their population by consuming organic mass as feed in the presence of oxygen. Generally, biochemical oxygen demand (BOD) used as measuring parameter for wastewater, as organic matters are removed, BOD level decreases [43]. This biological unit processes are further classified into suspended growth system and attached growth system.

In the suspended growth system, as its name suggest, the wastewater flows around and through the free-floating microorganisms, gathering as biological flocs, which is settled at the bottom. The settled flocs contain microorganisms which are recycled back to treat the wastewater. Activated sludge processes and aerated lagoons are the example of suspended growth systems. On the other hand, attached-growth process use media like fixed bed of gravel, ceramic, or plastic media to retain and grow the microorganisms. The wastewater flows over the media and create a biofilms that becomes thick on growth and falls or detached off as sloughing. Trickling filter and rotating biological contactors (RBC) are the example of the same.

Attached-growth processes are simpler operation, need less equipment maintenance and less energy as compared to suspended-growth system. At the same time, it required larger space, has odor issues and limitations to handle high volume of wastewater.

2.2.1 Aerobic attached growth system

In this system, the surface of the sewage is exposed to oxygen after adding a certain type of microorganism (aerobes) such as bacteria [44, 45]. These aerobes are oxidized the organic matter whichever liable to decay by using the oxygen. After the oxidation of the organic matter, a bacterial film or slime layer on the sewage surface area is formed. The organisms present in the film adsorb more of the organic matter and form as coagulated matter, consequently it will settle down. There are different types of methods available based on this, but mostly trickling filters and rotating biological contactors have been used.

2.2.1.1 Trickling filters

The trickling filters consist of a cylindrical tank and filled with high specific surface area materials including slag, coke, gravel ceramic, polyurethane foam, etc. (Fig. 4a) [46-48].

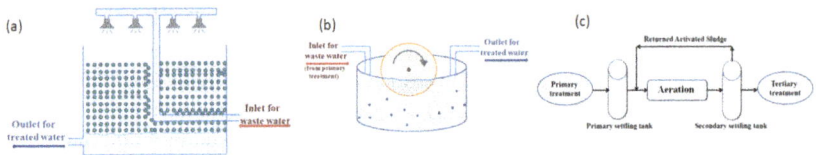

Figure 4. Secondary treatment process (a) Trickling filters; (b) Rotating biological contactor and (c) Flow Diagram of Activated Sludge Process.

A high specific surface provides a large area for biofilm formation. The aerobic bacteria grow in the thin film over the surface of the materials. After passing the wastewater from the bottom of the trickling filter, the water is sprayed into the bed through a rotating arm. The influent is then passed through the bacteria bed and comes out as effluent. For a better treatment of this organic matter, it is again re-circulated through the bed. It oxidizes the organic compounds in the wastewater to carbon dioxide and water.

2.2.1.2 Rotating biological contactors (RBC)

The rotating biological contactor unit contains a series of closely spaced polymeric circular discs mounted on a rotating shaft which is supported just above the surface of the wastewater [49, 50]. Microorganisms are grown on the surface of the circular discs, which is arranged on a horizontal shaft in a cylindrical tank and 40% submerged in wastewater (Fig. 4b). The disc rotates in and out of the wastewater as RBC rotates. The microorganisms gain the required oxygen when they are out of the water and the discs absorb organic compounds while they are inside the wastewater. This process leads to the removal of a large number of organic compounds.

2.2.2 Aerobic suspended growth systems

In suspended growth system aerobes are suspended in liquid medium with continuous mixing; which is responsible for the treatment [51]. Generally, these systems are two types based on whether the sludge is re-circulated or not. In this, the microorganisms convert the organic matter presents in the sewage into gases and cell tissue *via* metabolizing the organic matter by a microorganism. Currently, various methods developed based on this principle such as activated sludge process, aerated lagoons, and oxidation pond.

2.2.2.1 Activated sludge process

The effluent from the primary clarifier passes to the aeration tank. Meanwhile, this tank receives partially microorganisms from secondary clarifier or activated sludge (Fig. 4c) [51]. Subsequently, aeration takes place for maintaining the aerobic conditions for the oxidation of impurities. After several hours of agitation, the treated water sends to secondary clarifier for the settling of solids matter (sludge).

2.2.2.2 Aerated lagoons

After primary treatment, the wastewater enters into the lagoons to remove organic matter [52, 53]. Lagoons are fixed with mechanical aerators to oxidize the organic matter by using suspended aerobic microorganisms. The treated water sends for further treatment. Aerated lagoons are of two types; namely facultative aerated lagoons and aerobic lagoons. In the facultative aerated lagoon, solid matter settles down at the bottom layer and leads to anaerobic digestion. Subsequently, aerobic digestion takes place at the upper layer. But in the aerobic lagoons process, the wastewater is continuously aerated from top to bottom. It ceases the formation of a solid settlement at the bottom.

2.2.2.3 Oxidation pond

Oxidation pond constructed for the treatment of wastewater *via* interaction of bacteria, algae, and sunlight [54, 55]. In this strategy, various inorganic nutrients (nitrogen, phosphorus) and carbon-dioxide is released due to organic matter is metabolized by bacteria. These compounds are utilized by algae in the presence of sunlight and produce oxygen. This oxygen is taken up by bacteria and decreases the BOD through closing the cycle.

2.3 Tertiary treatment

Tertiary treatments are also named as advance treatment methods [56]. This step specifically removes a remarkable quantity of phosphorous, nitrogen, biodegradable organic matter, heavy metals, virus and pathogenic bacteria [57, 58]. There are various advanced treatment methods have been developed including disinfection, membrane separation, electrodialysis. The advantages and disadvantages of conventional waste water treatments along with tertiary treatment have been reported in table 2.

Table 2. Advantages and disadvantages of various wastewater treatment processes.

Technology	Advantages	Disadvantages
Coagulation	• Low principal cost • Effortless procedure	• Generating a large quantity of sludge
Trickling filters	• Capability to remove the high quantity of the BOD • Efficiency in treating a large amount of organic matter • Remarkable ability to remove ammonia • Creates a low amount of sludge • High skilled and technical expertise is not required to operate and manage the system	• High principal cost • Blocking of rotating arms • High intense odour generation
Rotating biological contactors	• Aeration *via* atmosphere • Affordable operation cost • Generation of low quantity of sludge	• A huge area is necessary for this system • Protection of the system in a cold climate is difficult
Activated sludge process	• Aeration *via* atmosphere • Available in domestic to industrial scale • Feasible to remove the high quantity of the BOD, COD, nitrogen, and phosphorous. • Simple operating system • Ability in producing electric energy from biogas • Excellent effluent quality	• Concentration the of sludge to be monitor • Operation cost
Disinfection (Bleaching powder/Cl_2 treatment)	• Low cost and readily available • High water solubility • Toxic to pathogens • Potential to remove a high quantity of the iron, manganese and ammonia nitrogen during	• Corrosive • Difficult to handle Cl_2 gas due to its hazards nature

	oxidation	
Reverse osmosis	• Available in domestic to industrial scale • The efficient rejection rate of salt, organic compounds, and pathogens. • The low quantity of energy is sufficient • Membranes can be replaced easily • Easy to maintain	• The problem in large scale treatment due to sludge generation • Pre-treatment is essential before purifying the water • Membranes are highly sensitive to pH
Electrodialysis	• High selectivity of ions • Regeneration of membrane is easy • Pre-treatment of water not necessary	• Operational cost is high • Ions only can be removed • Based on water quality, the membrane needs to be selective
UASBR	• Energy demand, land requirement, and sludge production is low • Economical • Excellent organic and BOD removal efficiency • Effluents can be used for farming	• Extended stand-up period • Requires enough quantity of granular seed sludge • Lesser gas yield

2.3.1 Disinfection process

Disinfection process method mostly used for the treatment of municipal water at the tertiary level [59-62]. The process of destroying the pathogenic microorganisms including bacteria and viruses from the water using certain chemicals (chlorine; disinfectant) is called disinfection. The disinfection of municipal water can be carried out by chlorination method viz. either adding bleaching powder or purging chlorine gas.

In the bleaching power treatment, the required quantity of bleaching powder is added to municipal wastewater and kept for many hours without disturbing [63]. During this process, bleaching powder is reacted with water and forms chlorine gas (Fig. 5). Further, this chlorine (Cl_2) gas is reacted with water and forms hypo-chlorous acid (HOCl), it acts as a germicide. Besides this chlorination process instead of adding bleaching powder, chlorine gas is directly purged into municipal water form HOCl germicide [64, 65]. This germicide destroys the pathogenic microorganisms.

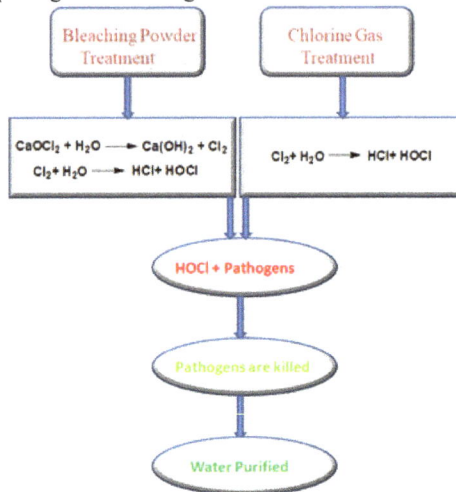

Figure 5. Flow chart for disinfection action.

2.3.2 Reverse osmosis

Reverse osmosis (RO) used to remove inorganic minerals (calcium, magnesium, sodium, potassium, phosphorous, and fluoride) and organic compounds which include pesticides [66-68].

In the RO process the wastewater passes through a semi-permeable membrane by applying osmotic pressure (Fig. 6a). RO membranes contain tiny pores through which water can flow [69, 70]. These pores are not only capable to restrict impurities but also allow the water to pass through. So, in RO treatment membranes are playing a key role in the purification of wastewater. The most commonly used membranes are cellulose acetate and polyamide.

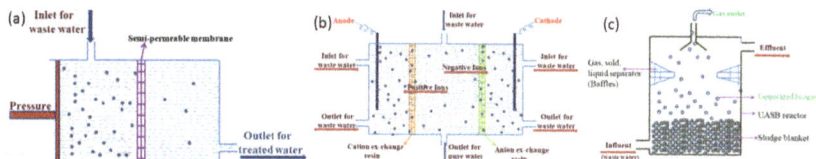

Figure 6. Tertiary treatment process (a) Reverse Osmosis Process; (b) Electrodialysis; (c) Up-flow anaerobic sludge blanket reactor.

2.3.3 Electrodialysis

The electrodialysis process is used to remove cations and anions present in water [71, 72]. In this process 'ions' present in water migrates through ion-exchange membranes under the influence of applied potential (Fig. 6b).

A single Electrodialysis unit contains a cation-selective membrane and an anion-selective membrane [73]. These membranes placed between the cathode (-ve) and anode (+ve) electrodes. During the process, cations will migrate towards cathode whereas anions migrate towards anode upon applying an external potential. In this process, cation-selective membranes allow cations and anion-selective membranes to allow anions, consequently, the water gets purified.

2.3.4 Up-flow anaerobic sludge blanket reactor (UASBR)

In this process, the wastewater passes through the bottom of the UASBR reactor (Fig 6c), which contains granular anaerobic biomass [74-76]. The influent is pumped from the bottom of the UASBR through pump. The biomass in the sludge bed helps in the conversion of sludge present in wastewater to different gases through hydrolysis, acetogenesis, and methanogenesis processes. Various gases (H_2, CO_2, and CH_4) are generated in this process, which are collected at the top of the reactor. Top of the reactor fixed with a three-phase separator for the separation of biogas produced in the reactor. Especially, this process removes chemical oxygen demand (COD) up to 80% [75].

3. Advance water purification technologies

Even though various kinds of traditional technologies have been recognized for domestic and industrial water treatment, these methods are limited to a certain level [77-86]. So, removing pollutants like various metals, organic and inorganic compounds, and pathogens from water are becoming a difficult task. Therefore, researchers came with various new techniques by using materials like activated carbon [87-97], nanomaterials [98-106], ozonation [107-113], and ultraviolet radiation [114-118] to enhance the pollutant removal capability. Here the discussion is based on the above materials and their analog technique.

3.1   Activated carbon filters

Activated carbon-based technology has been developed as a water filtering medium for purification of both domestic and industrial water [119-124]. Activated carbon can adsorb pollutants due to its high surface area composed of porous material [125-128]. Thus, this method enhances the adsorption capability of pollutants on the surface of the activated carbon material. Additionally, the water further interacted with different sized sand particles i.e. fine sand, coarse sand, and gravels to improve the purification. Fig 7a shows the schematic illustration of the activated carbon filter assisted water purification system. These methods mainly involve three basic steps

1. Pollutants interact and adsorb to peripheral of the carbon material

2. Pollutants move to into carbon pores

3. Pollutants adsorb on the internal walls of the carbon

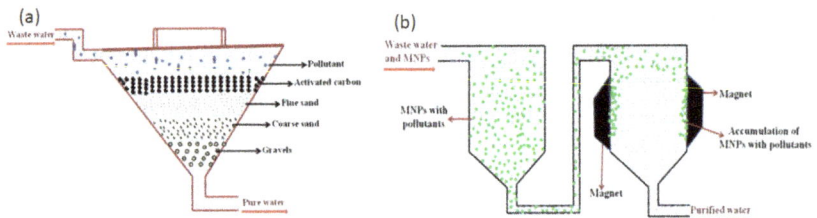

Figure 7. Illustration of (a) activated carbon filter assisted water purification system; (b) activated MNPs based water purification system.

3.2   Magnetic Nanoparticles

The development of nanoscience and nanotechnology creates a nano-platform for various applications including analytical [129-131], bio-medical [132-138], and environmental [139-144]. The distinctive size-related features of nanoscale materials have a remarkable impact on nanotechnology and making a promising platform for industrial and domestic water treatment [145-147].

Among the various kinds of nanomaterials, magnetic natured nanoparticles have been found not only the easiest and cost-effective way in purifying the wastewater but also its recovery after purification by a simple magnet. In the category, Fe_3O_4 magnetic nanoparticles (MNPs) showed a promising path for the removal of metals as well as other

Materials Research Forum LLC
https://doi.org/10.21741/9781644901144-1

pollutants [148-152]. Fig 7b represents the schematic illustration of the MNPs based water purification system. Like activated carbon, these MNPs have also shown its adsorption behaviour towards the pollutants present in water [153, 154]. MNPs based adsorbents are showing better adsorption to pollutants due to their high surface to volume ratio as well as magnetic behaviour. Further, the surface modification of MNPs by using various organic and biomolecules enhances the adsorption ability of various toxic pollutants from the contaminated water. Also, its reusability and the simple re-activation method after the removal of pollutants from water enhances the interest to researchers towards this area.

3.3   Ozonation

Ozonation is a type of advanced oxidation process (Fig 8a), involving the generation of highly reactive oxygen species (ROS) [155-160]. Ozone is an unstable gas encompassing three oxygen atoms, the gas will rapidly degrade and converted to oxygen [161]. During this alteration ROS form such as a free oxygen atom, or free radical. These ROS are highly reactive and short-lived, under normal conditions. Additionally, ROS are capable to attack not only a variety of pathogens but also organic and inorganic pollutants. Thus, the treatment of water with ozone is efficient for disinfection as well as for the degradation of organic and inorganic pollutants. The main advantage of this method is its simple operation as well as the absence of sludge production.

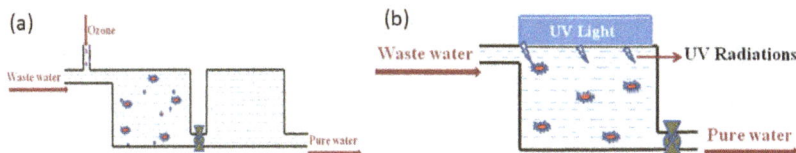

Figure 8. (a) Ozonation process; (b) UV treatment process for disinfection.

3.4   Ultraviolet (UV) radiation

The treatment of water with UV radiation is efficient for disinfection (Fig 8b) [162-181]. Presently, emerging water treatment technology equipped with UV radiation and applying in both domestic and industrial wastewater. In this method, UV radiation penetrates the cell wall of the micro-organism and damages the genetic mutant [182, 183]. Consequently, it prevents the reproduction of cells. Inappropriate quantities, UV radiation will destroy nearly 99.9% of micro-organisms present within a water stream.

3.5   Lime-Soda (LS) process

Lime-soda water softening strategy is one of the significant chemical procedure for decreasing the hardness of the water [184-188]. In this process required quantities of lime (CaO) and soda (Na_2CO_3) are added into the reaction tank containing hard water along with the super-heated steam (Figure 9a). Under high-temperature condition of steam, the chemicals are reacted with hardness causing salts present in water. Simultaneously, the reacted salts are precipitated as carbonates and hydroxides in the form of sludge into a sedimentation tank. The formed sludge can be collected from the sludge outlet. Besides this, the treated water is sent into the sand filter which is attached to the sedimentation tank. During this, the hardness removed water interacted with different sized sand particles which are present in a sand filter, to remove the suspended sludge particles.

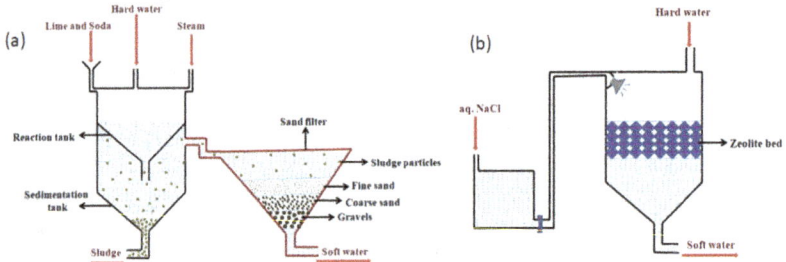

Figure 9. (a) Lime Soda process for softening water; (b) Zeolite process for softening water.

3.6   Zeolite process

Zeolites are hydrated sodium alumina silicates, having the capability to remove the hardness from the water *via* sodium ions exchange [189-194]. In this approach, the hard water is passed through the zeolite bed. During this process, the hardness causing ions has interacted with the bed and exchange with sodium ions into the water, it leads to softening of water (figure 9b). The hardness removing capability of zeolite will be ceased at a certain point due to the loss of sodium ions from the zeolite bed. Therefore, the zeolite bed is treated with sodium chloride solution to regenerate the sodium ions containing the zeolite bed. By this process, the softening of water takes place up to a value of 10 ppm.

4.   Future wastewater treatment technologies

As discussed in section 1 and 2, conventional wastewater treatment technologies mainly consist of primary treatment based on adding chemicals/flocculants, secondary treatment by means of biological methods and polishing treatment based on adsorption through activated carbon. At present, conventional wastewater treatment technologies are facing so many problems in terms of higher rate of chemical consumptions, sludge disposal, higher energy and space requirements. Moreover, efficient removal of complex organic compounds, limitation of handling more wastewater than the fixed design capacity and lack of skilled manpower are also big operational issued in these technologies.

Due to all these operational and technological barriers in conventional wastewater treatment technologies, researchers are trying to develop new classes of advance wastewater treatment technologies to overcome the above mentioned problems. Advanced wastewater technology require to incorporate (i) Membrane technology (ii) Advanced Oxidation Processes (iii) Less amount of Sludge generation and even if sludge is generate than how to utilize the sludge rather than disposal at land fill site (iv) Low cost adsorption materails (v) Less amount of chemical consumptions or bioflocculants (vi) New class of nanomaterials for the treatment of wastewater. There is a huge quantum of research work available on the above mentioned points, still research gaps exist in open literature to address the issues of advances in wastewater treatment technologies. The utilization of advanced wastewater technologies in combination with conventional technologies may resulted into effective wastewater treatment and increase in reuse and recycle of treated water.

Conclusion

Water is part of our life as we cannot survive a single day without it. Most of the available water becomes contaminated and changing its physical, chemical, and biological nature due to various human activities. To ensure a harmless environment and public health, it must be purified. There are various conventional methods for wastewater treatment. The selection of a specific method mainly relies on the characteristics of wastewater whether it is from a municipality or industry. Sometimes it is also based on cost implications, technical expertise for operation and maintenance.

Here in this chapter, we have discussed various stages of wastewater treatments such as primary, secondary, and tertiary. Further, we have discussed various physical, chemical, and biological conventional methods involved in the wastewater treatments. Moreover, a brief summary of advanced treatment methods have been reported along with glimpse of future research in the field of wastewater treatment.

Acknowledgements

Balakrishna Kolli is thankful for the DST-SERB for the funding (Project No. ECR/2016/001216).

References

[1] E. Jéquier, F. Constant, Water as an essential nutrient: the physiological basis of hydration, European journal of clinical nutrition 64(2) (2010) 115-123. https://doi.org/10.1038/ejcn.2009.111

[2] S. Postel, K. Bawa, L. Kaufman, C.H. Peterson, S. Carpenter, D. Tillman, P. Dayton, S. Alexander, K. Lagerquist, L. Goulder, Nature's services: Societal dependence on natural ecosystems, Island Press2012.

[3] H. WHO, World Health Organization guidelines for drinking water quality, World Health Organisation Publication, New York (2008) 197-209.

[4] Y.J. Chan, M.F. Chong, C.L. Law, D. Hassell, A review on anaerobic–aerobic treatment of industrial and municipal wastewater, Chemical Engineering Journal 155(1-2) (2009) 1-18. https://doi.org/10.1016/j.cej.2009.06.041

[5] R. Nugent, The impact of urban agriculture on the household and local economies, Bakker N., Dubbeling M., Gündel S., Sabel-Koshella U., de Zeeuw H. Growing cities, growing food. Urban agriculture on the policy agenda. Feldafing, Germany: Zentralstelle für Ernährung und Landwirtschaft (ZEL) (2000) 67-95.

[6] E. El-Bestawy, Treatment of mixed domestic–industrial wastewater using cyanobacteria, Journal of industrial microbiology & biotechnology 35(11) (2008) 1503-1516. https://doi.org/10.1007/s10295-008-0452-4

[7] S. Ekino, M. Susa, T. Ninomiya, K. Imamura, T. Kitamura, Minamata disease revisited: an update on the acute and chronic manifestations of methyl mercury poisoning, Journal of the neurological sciences 262(1-2) (2007) 131-144. https://doi.org/10.1016/j.jns.2007.06.036

[8] S. Satarug, S.H. Garrett, M.A. Sens, D.A. Sens, Cadmium, environmental exposure, and health outcomes, Environmental health perspectives 118(2) (2010) 182-190. https://doi.org/10.1289/ehp.0901234

[9] D. Majumdar, The blue baby syndrome, Resonance 8(10) (2003) 20-30. https://doi.org/10.1007/BF02840703

Materials Research Forum LLC
https://doi.org/10.21741/9781644901144-1

[10] N.S. Bryan, D.D. Alexander, J.R. Coughlin, A.L. Milkowski, P. Boffetta, Ingested nitrate and nitrite and stomach cancer risk: an updated review, Food and Chemical Toxicology 50(10) (2012) 3646-3665. https://doi.org/10.1016/j.fct.2012.07.062

[11] H. Horowitz, Fluoride and enamel defects, Advances in dental research 3(2) (1989) 143-146. https://doi.org/10.1177/08959374890030021201

[12] G.-J. Zhou, G.-G. Ying, S. Liu, L.-J. Zhou, Z.-F. Chen, F.-Q. Peng, Simultaneous removal of inorganic and organic compounds in wastewater by freshwater green microalgae, Environmental Science: Processes & Impacts 16(8) (2014) 2018-2027. https://doi.org/10.1039/C4EM00094C

[13] A. Sonune, R. Ghate, Developments in wastewater treatment methods, Desalination 167 (2004) 55-63. https://doi.org/10.1016/j.desal.2004.06.113

[14] D. Mohan, A. Sarswat, Y.S. Ok, C.U. Pittman Jr, Organic and inorganic contaminants removal from water with biochar, a renewable, low cost and sustainable adsorbent–a critical review, Bioresource technology 160 (2014) 191-202. https://doi.org/10.1016/j.biortech.2014.01.120

[15] G. Gedda, C.-Y. Lee, Y.-C. Lin, H.-f. Wu, Green synthesis of carbon dots from prawn shells for highly selective and sensitive detection of copper ions, Sensors and Actuators B: Chemical 224 (2016) 396-403. https://doi.org/10.1016/j.snb.2015.09.065

[16] G. Gedda, H.N. Abdelhamid, M.S. Khan, H.-F. Wu, ZnO nanoparticle-modified polymethyl methacrylate-assisted dispersive liquid–liquid microextraction coupled with MALDI-MS for rapid pathogenic bacteria analysis, RSC advances 4(86) (2014) 45973-45983. https://doi.org/10.1039/C4RA03391D

[17] M.L. Bhaisare, G. Gedda, M.S. Khan, H.-F. Wu, Fluorimetric detection of pathogenic bacteria using magnetic carbon dots, Analytica chimica acta 920 (2016) 63-71. https://doi.org/10.1016/j.aca.2016.02.025

[18] G. Gedda, S. Pandey, Y.-C. Lin, H.-F. Wu, Antibacterial effect of calcium oxide nano-plates fabricated from shrimp shells, Green Chemistry 17(6) (2015) 3276-3280. https://doi.org/10.1039/C5GC00615E

[19] T. Reemtsma, U. Miehe, U. Duennbier, M. Jekel, Polar pollutants in municipal wastewater and the water cycle: occurrence and removal of benzotriazoles, Water research 44(2) (2010) 596-604. https://doi.org/10.1016/j.watres.2009.07.016

[20] M. Panizza, G. Cerisola, Removal of organic pollutants from industrial wastewater by electrogenerated Fenton's reagent, Water Research 35(16) (2001) 3987-3992. https://doi.org/10.1016/S0043-1354(01)00135-X

Materials Research Forum LLC
https://doi.org/10.21741/9781644901144-1

[21] C. Wang, X. Hu, M.-L. Chen, Y.-H. Wu, Total concentrations and fractions of Cd, Cr, Pb, Cu, Ni and Zn in sewage sludge from municipal and industrial wastewater treatment plants, Journal of hazardous materials 119(1-3) (2005) 245-249. https://doi.org/10.1016/j.jhazmat.2004.11.023

[22] H. Brix, Wastewater treatment in constructed wetlands: system design, removal processes, and treatment performance, Constructed wetlands for water quality improvement (1993) 9-22. https://doi.org/10.1201/9781003069997-3

[23] V.K. Gupta, I. Ali, T.A. Saleh, A. Nayak, S. Agarwal, Chemical treatment technologies for waste-water recycling—an overview, Rsc Advances 2(16) (2012) 6380-6388. https://doi.org/10.1039/c2ra20340e

[24] M.R. Templeton, D. Butler, Introduction to wastewater treatment, Bookboon2011.

[25] F.R. Spellman, Handbook of water and wastewater treatment plant operations, CRC press2013. https://doi.org/10.1201/b15579

[26] T.M. Pankratz, Screening Equipment Handbook, CRC Press1995.

[27] J.P. Guyer, An Introduction to Preliminary Wastewater Treatment, Guyer Partners2018.

[28] F.R. Spellman, Water and wastewater treatment plant operations, New York 2003. 17 Iurciuc CE Teză de doctorat. Studii şi cercetări privind epurarea avansată a apelor uzate în vederea valorificării efluentului la irigarea culturilor energetice. Iaşi 2013 (2009). https://doi.org/10.1201/9780203489833

[29] G. Balkiş, Experimental investigation of energy dissipation through inclined screens, M. Sc. Thesis, Department of Civil Engineering Middle East Technical …, 2004.

[30] L. Hernesniemi, Screen having inclined slots for a digester, Google Patents, 2002.

[31] L. Hernesniemi, Screen having inclined slots for use in a continuous digester, Google Patents, 2000.

[32] T.R. Camp, Grit chamber design, Sewage Works Journal (1942) 368-381.

[33] O.E. Albertson, Aerated grit chamber and method, Google Patents, 2005.

[34] G. Karia, R. Christian, Wastewater treatment: Concepts and design approach, PHI Learning Pvt. Ltd.2013.

[35] M. Wason, S. Purhoit, D. Dehon, M. Magee, The wastewater treatment process, BE, 2007.

[36] H. Pentz, C. Parkhani, F. Majeron, Skimming apparatus for clarification tank, Google Patents, 1973.

[37] A. Razmi, B. FIROUZABADI, G. Ahmadi, Experimental and numerical approach to enlargement of performance of primary settling tanks, (2009).

[38] F. Chagnon, D.R. Harleman, An Introduction to Chemically Enhanced Primary Treatment, Massachusetts Institute of Technology, Cambridge, MA (2002).

[39] S.A. Al-Jlil, COD and BOD reduction of domestic wastewater using activated sludge, sand filters and activated carbon in Saudi Arabia, Biotechnology 8(4) (2009) 473-477. https://doi.org/10.3923/biotech.2009.473.477

[40] K. Sivagami, K. Sakthivel, I.M. Nambi, Advanced oxidation processes for the treatment of tannery wastewater, Journal of environmental chemical engineering 6(3) (2018) 3656-3663. https://doi.org/10.1016/j.jece.2017.06.004

[41] D.E. Severeid, D.D. Jech, Process for the secondary treatment of wastewater, Google Patents, 1983.

[42] T.S. Tsang, J.B. Seward, M.E. Barnes, K.R. Bailey, L.J. Sinak, L.H. Urban, S.N. Hayes, Outcomes of primary and secondary treatment of pericardial effusion in patients with malignancy, Mayo Clinic Proceedings, Elsevier, 2000, pp. 248-253. https://doi.org/10.4065/75.3.248

[43] V.V. Ranade, V.M. Bhandari, Industrial wastewater treatment, recycling and reuse, Butterworth-Heinemann2014. https://doi.org/10.1016/B978-0-08-099968-5.00014-3

[44] G. Parkin, R. Speece, Attached versus suspended growth anaerobic reactors: response to toxic substances, Water Science and Technology 15(8-9) (1983) 261-289. https://doi.org/10.2166/wst.1983.0171

[45] T.I. Tatoulis, A.G. Tekerlekopoulou, C.S. Akratos, S. Pavlou, D.V. Vayenas, Aerobic biological treatment of second cheese whey in suspended and attached growth reactors, Journal of Chemical Technology & Biotechnology 90(11) (2015) 2040-2049. https://doi.org/10.1002/jctb.4515

[46] E. Eding, A. Kamstra, J. Verreth, E. Huisman, A. Klapwijk, Design and operation of nitrifying trickling filters in recirculating aquaculture: a review, Aquacultural engineering 34(3) (2006) 234-260. https://doi.org/10.1016/j.aquaeng.2005.09.007

[47] O.-I. Lekang, H. Kleppe, Efficiency of nitrification in trickling filters using different filter media, Aquacultural engineering 21(3) (2000) 181-199.

[48] T. Wik, Trickling filters and biofilm reactor modelling, Reviews in Environmental Science and Biotechnology 2(2-4) (2003) 193-212. https://doi.org/10.1016/S0144-8609(99)00032-1

[49] F. Hassard, J. Biddle, E. Cartmell, B. Jefferson, S. Tyrrel, T. Stephenson, Rotating biological contactors for wastewater treatment–a review, Process Safety and Environmental Protection 94 (2015) 285-306. https://doi.org/10.1016/j.psep.2014.07.003

[50] S. Cortez, P. Teixeira, R. Oliveira, M. Mota, Rotating biological contactors: a review on main factors affecting performance, Reviews in Environmental Science and Bio/Technology 7(2) (2008) 155-172. https://doi.org/10.1007/s11157-008-9127-x

[51] S.V. Mohan, N.C. Rao, K.K. Prasad, B. Madhavi, P. Sharma, Treatment of complex chemical wastewater in a sequencing batch reactor (SBR) with an aerobic suspended growth configuration, Process Biochemistry 40(5) (2005) 1501-1508. https://doi.org/10.1016/j.procbio.2003.02.001

[52] R. Chandra, R.N. Bharagava, A. Kapley, H.J. Purohit, Bacterial diversity, organic pollutants and their metabolites in two aeration lagoons of common effluent treatment plant (CETP) during the degradation and detoxification of tannery wastewater, Bioresource technology 102(3) (2011) 2333-2341. https://doi.org/10.1016/j.biortech.2010.10.087

[53] A. Moura, M. Tacao, I. Henriques, J. Dias, P. Ferreira, A. Correia, Characterization of bacterial diversity in two aerated lagoons of a wastewater treatment plant using PCR–DGGE analysis, Microbiological Research 164(5) (2009) 560-569. https://doi.org/10.1016/j.micres.2007.06.005

[54] E. Butler, Y.-T. Hung, M.S. Al Ahmad, R.Y.-L. Yeh, R.L.-H. Liu, Y.-P. Fu, Oxidation pond for municipal wastewater treatment, Applied water science 7(1) (2017) 31-51. https://doi.org/10.1007/s13201-015-0285-z

[55] A. Ogunfowokan, E. Okoh, A. Adenuga, O. Asubiojo, An assessment of the impact of point source pollution from a university sewage treatment oxidation pond on a receiving stream–a preliminary study, Journal of applied sciences 5(1) (2005) 36-43. https://doi.org/10.3923/jas.2005.36.43

[56] T. Serra, J. Colomer, C. Pau, M. Marín, L. Sala, Tertiary treatment for wastewater reuse based on the Daphnia magna filtration–comparison with conventional tertiary treatments, Water science and technology 70(4) (2014) 705-711. https://doi.org/10.2166/wst.2014.284

[57] B. Chen, Y. Kim, P. Westerhoff, Occurrence and treatment of wastewater-derived organic nitrogen, Water research 45(15) (2011) 4641-4650. https://doi.org/10.1016/j.watres.2011.06.018

[58] N.M. Al-Bastaki, Performance of advanced methods for treatment of wastewater: UV/TiO2, RO and UF, Chemical Engineering and Processing: Process Intensification 43(7) (2004) 935-940. https://doi.org/10.1016/j.cep.2003.08.003

[59] B.A. Lyon, A.D. Dotson, K.G. Linden, H.S. Weinberg, The effect of inorganic precursors on disinfection byproduct formation during UV-chlorine/chloramine drinking water treatment, Water research 46(15) (2012) 4653-4664. https://doi.org/10.1016/j.watres.2012.06.011

[60] P.K. Roy, D. Kumar, M. Ghosh, A. Majumder, Disinfection of water by various techniques–comparison based on experimental investigations, Desalination and water treatment 57(58) (2016) 28141-28150. https://doi.org/10.1080/19443994.2016.1183522

[61] A. Pant, A.K. Mittal, Disinfection of wastewater: Comparative evaluation of chlorination and DHS-biotower, Journal of environmental biology 28(4) (2007) 717.

[62] J.G. Jacangelo, R.R. Trussell, International report: Water and wastewater disinfection-trends, issues and practices, Water science and technology: water supply 2(3) (2002) 147-157. https://doi.org/10.2166/ws.2002.0097

[63] D. Mazumder, Process evaluation and treatability study of wastewater in a textile dyeing industry, International Journal of Energy & Environment 2(6) (2011).

[64] S.W.E. Disinfection, Disinfection of wastewater effluent: Comparison of alternative technologies, (2008).

[65] M.A. Brown, S. Miller, G.L. Emmert, On-line purge and trap gas chromatography for monitoring of trihalomethanes in drinking water distribution systems, Analytica chimica acta 592(2) (2007) 154-161. https://doi.org/10.1016/j.aca.2007.04.020

[66] C.A. Dyke, Process for removing organic and inorganic contaminants from refinery wastewater streams employing ultrafiltration and reverse osmosis, Google Patents, 2000.

[67] R.Y. Ning, Arsenic removal by reverse osmosis, Desalination 143(3) (2002) 237-241. https://doi.org/10.1016/S0011-9164(02)00262-X

[68] L.L. Dueker, C.B. Cluff, Method and apparatus for removing minerals from a water source, Google Patents, 2008.

[69] S. Yüksel, N. Kabay, M. Yüksel, Removal of bisphenol A (BPA) from water by various nanofiltration (NF) and reverse osmosis (RO) membranes, Journal of hazardous materials 263 (2013) 307-310. https://doi.org/10.1016/j.jhazmat.2013.05.020

[70] S.-Y. Kwak, S.G. Jung, S.H. Kim, Structure-motion-performance relationship of flux-enhanced reverse osmosis (RO) membranes composed of aromatic polyamide thin films, Environmental science & technology 35(21) (2001) 4334-4340.

[71] H. Strathmann, Electrodialysis, a mature technology with a multitude of new applications, Desalination 264(3) (2010) 268-288. https://doi.org/10.1016/j.desal.2010.04.069

[72] H. AlMadani, Water desalination by solar powered electrodialysis process, Renewable Energy 28(12) (2003) 1915-1924. https://doi.org/10.1016/S0960-1481(03)00014-4

[73] X. Tongwen, Electrodialysis processes with bipolar membranes (EDBM) in environmental protection—a review, Resources, conservation and recycling 37(1) (2002) 1-22. https://doi.org/10.1016/S0921-3449(02)00032-0

[74] M.A. Latif, R. Ghufran, Z.A. Wahid, A. Ahmad, Integrated application of upflow anaerobic sludge blanket reactor for the treatment of wastewaters, Water research 45(16) (2011) 4683-4699. https://doi.org/10.1016/j.watres.2011.05.049

[75] H.N. Gavala, H. Kopsinis, I. Skiadas, K. Stamatelatou, G. Lyberatos, Treatment of dairy wastewater using an upflow anaerobic sludge blanket reactor, Journal of Agricultural Engineering Research 73(1) (1999) 59-63. https://doi.org/10.1006/jaer.1998.0391

[76] Y. Liu, H.-L. Xu, S.-F. Yang, J.-H. Tay, Mechanisms and models for anaerobic granulation in upflow anaerobic sludge blanket reactor, Water Research 37(3) (2003) 661-673. https://doi.org/10.1016/S0043-1354(02)00351-2

[77] I. Oller, S. Malato, J. Sánchez-Pérez, Combination of advanced oxidation processes and biological treatments for wastewater decontamination—a review, Science of the total environment 409(20) (2011) 4141-4166. https://doi.org/10.1016/j.scitotenv.2010.08.061

[78] M.I. Gil, M.V. Selma, F. López-Gálvez, A. Allende, Fresh-cut product sanitation and wash water disinfection: problems and solutions, International journal of food microbiology 134(1-2) (2009) 37-45. https://doi.org/10.1016/j.ijfoodmicro.2009.05.021

[79] M. Gavrilescu, Removal of heavy metals from the environment by biosorption, Engineering in Life Sciences 4(3) (2004) 219-232. https://doi.org/10.1002/elsc.200420026

[80] M. Fomina, G.M. Gadd, Biosorption: current perspectives on concept, definition and application, Bioresource technology 160 (2014) 3-14. https://doi.org/10.1016/j.biortech.2013.12.102

[81] P.H. Gleick, M. Palaniappan, Peak water limits to freshwater withdrawal and use, Proceedings of the National Academy of Sciences 107(25) (2010) 11155-11162. https://doi.org/10.1073/pnas.1004812107

[82] A. Babuponnusami, K. Muthukumar, A review on Fenton and improvements to the Fenton process for wastewater treatment, Journal of Environmental Chemical Engineering 2(1) (2014) 557-572. https://doi.org/10.1016/j.jece.2013.10.011

[83] S. Madaeni, The application of membrane technology for water disinfection, Water Research 33(2) (1999) 301-308. https://doi.org/10.1016/S0043-1354(98)00212-7

[84] I. Ali, New generation adsorbents for water treatment, Chemical reviews 112(10) (2012) 5073-5091. https://doi.org/10.1021/cr300133d

[85] M.S. Mohsen, O.R. Al-Jayyousi, Brackish water desalination: an alternative for water supply enhancement in Jordan, Desalination 124(1) (1999) 163-174. https://doi.org/10.1016/S0011-9164(99)00101-0

[86] J. Mallevialle, P.E. Odendaal, M.R. Wiesner, Water treatment membrane processes, American Water Works Association1996.

[87] A. Bhatnagar, W. Hogland, M. Marques, M. Sillanpää, An overview of the modification methods of activated carbon for its water treatment applications, Chemical Engineering Journal 219 (2013) 499-511. https://doi.org/10.1016/j.cej.2012.12.038

[88] F. Beltrán, F. Masa, J. Pocostales, A comparison between catalytic ozonation and activated carbon adsorption/ozone-regeneration processes for wastewater treatment, Applied Catalysis B: Environmental 92(3-4) (2009) 393-400. https://doi.org/10.1016/j.apcatb.2009.08.019

[89] D. Kalderis, S. Bethanis, P. Paraskeva, E. Diamadopoulos, Production of activated carbon from bagasse and rice husk by a single-stage chemical activation method at low retention times, Bioresource technology 99(15) (2008) 6809-6816.

[90] V.K. Gupta, A. Mittal, R. Jain, M. Mathur, S. Sikarwar, Adsorption of Safranin-T from wastewater using waste materials—activated carbon and activated rice husks,

Journal of Colloid and Interface Science 303(1) (2006) 80-86. https://doi.org/10.1016/j.jcis.2006.07.036

[91] R. Thiruvenkatachari, W.G. Shim, J.W. Lee, R.B. Aim, H. Moon, A novel method of powdered activated carbon (PAC) pre-coated microfiltration (MF) hollow fiber hybrid membrane for domestic wastewater treatment, Colloids and Surfaces A: Physicochemical and Engineering Aspects 274(1-3) (2006) 24-33. https://doi.org/10.1016/j.colsurfa.2005.08.026

[92] H. Zhang, Regeneration of exhausted activated carbon by electrochemical method, Chemical Engineering Journal 85(1) (2002) 81-85. https://doi.org/10.1016/S1385-8947(01)00176-0

[93] O. Amuda, A. Ibrahim, Industrial wastewater treatment using natural material as adsorbent, African Journal of Biotechnology 5(16) (2006).

[94] S. Wong, N. Ngadi, I.M. Inuwa, O. Hassan, Recent advances in applications of activated carbon from biowaste for wastewater treatment: a short review, Journal of Cleaner Production 175 (2018) 361-375. https://doi.org/10.1016/j.jclepro.2017.12.059

[95] J.R. Perrich, Activated carbon adsorption for wastewater treatment, CRC press2018. https://doi.org/10.1201/9781351069465

[96] F. Bonvin, L. Jost, L. Randin, E. Bonvin, T. Kohn, Super-fine powdered activated carbon (SPAC) for efficient removal of micropollutants from wastewater treatment plant effluent, Water research 90 (2016) 90-99. https://doi.org/10.1016/j.watres.2015.12.001

[97] H. Zeng, H. Hongming, G.E.M. Center, Optimization of Cadmium-containing Wastewater Treatment with Activated Carbon by Response Surface Method, Environmental Science and Management (1) (2018) 25.

[98] B. Bethi, S.H. Sonawane, B.A. Bhanvase, S.P. Gumfekar, Nanomaterials-based advanced oxidation processes for wastewater treatment: a review, Chemical Engineering and Processing-Process Intensification 109 (2016) 178-189. https://doi.org/10.1016/j.cep.2016.08.016

[99] T.A. Saleh, An Overview of Nanomaterials for Water Technology, Advanced Nanomaterials for Water Engineering, Treatment, and Hydraulics, IGI Global2017, pp. 1-12. https://doi.org/10.4018/978-1-5225-2136-5.ch001

[100] I.C. Butnariu, O. Stoian, Ş. Voicu, H. Iovu, G. Paraschiv, NANOMATERIALS USED IN TREATMENT OF WASTEWATER: A REVIEW, Annals of the Faculty of Engineering Hunedoara 17(2) (2019) 175-179.

Materials Research Forum LLC
https://doi.org/10.21741/9781644901144-1

[101] A.B. Holmes, F.X. Gu, Emerging nanomaterials for the application of selenium removal for wastewater treatment, Environmental Science: Nano 3(5) (2016) 982-996. https://doi.org/10.1039/C6EN00144K

[102] Y. Zhang, B. Wu, H. Xu, H. Liu, M. Wang, Y. He, B. Pan, Nanomaterials-enabled water and wastewater treatment, NanoImpact 3 (2016) 22-39. https://doi.org/10.1016/j.impact.2016.09.004

[103] R. Thines, N. Mubarak, S. Nizamuddin, J. Sahu, E. Abdullah, P. Ganesan, Application potential of carbon nanomaterials in water and wastewater treatment: a review, Journal of the Taiwan Institute of Chemical Engineers 72 (2017) 116-133. https://doi.org/10.1016/j.jtice.2017.01.018

[104] O. Suárez-Iglesias, S. Collado, P. Oulego, M. Díaz, Graphene-family nanomaterials in wastewater treatment plants, Chemical Engineering Journal 313 (2017) 121-135. https://doi.org/10.1016/j.cej.2016.12.022

[105] Z. Cai, Y. Sun, W. Liu, F. Pan, P. Sun, J. Fu, An overview of nanomaterials applied for removing dyes from wastewater, Environmental Science and Pollution Research 24(19) (2017) 15882-15904. https://doi.org/10.1007/s11356-017-9003-8

[106] H. Sadegh, G.A. Ali, Potential applications of nanomaterials in wastewater treatment: nanoadsorbents performance, Advanced Treatment Techniques for Industrial Wastewater, IGI Global2019, pp. 51-61. https://doi.org/10.4018/978-1-5225-5754-8.ch004

[107] M. Bourgin, B. Beck, M. Boehler, E. Borowska, J. Fleiner, E. Salhi, R. Teichler, U. Von Gunten, H. Siegrist, C.S. McArdell, Evaluation of a full-scale wastewater treatment plant upgraded with ozonation and biological post-treatments: Abatement of micropollutants, formation of transformation products and oxidation by-products, Water research 129 (2018) 486-498. https://doi.org/10.1016/j.watres.2017.10.036

[108] J.E. Schollée, M. Bourgin, U. von Gunten, C.S. McArdell, J. Hollender, Non-target screening to trace ozonation transformation products in a wastewater treatment train including different post-treatments, Water research 142 (2018) 267-278. https://doi.org/10.1016/j.watres.2018.05.045

[109] J.C. Cardoso, G.G. Bessegato, M.V.B. Zanoni, Efficiency comparison of ozonation, photolysis, photocatalysis and photoelectrocatalysis methods in real textile wastewater decolorization, Water research 98 (2016) 39-46. https://doi.org/10.1016/j.watres.2016.04.004

[110] K. Paździor, J. Wrębiak, A. Klepacz-Smółka, M. Gmurek, L. Bilińska, L. Kos, J. Sójka-Ledakowicz, S. Ledakowicz, Influence of ozonation and biodegradation on toxicity of industrial textile wastewater, Journal of Environmental Management 195 (2017) 166-173. https://doi.org/10.1016/j.jenvman.2016.06.055

[111] C. Von Sonntag, U. Von Gunten, Chemistry of ozone in water and wastewater treatment, IWA publishing2012. https://doi.org/10.2166/9781780400839

[112] L. Chu, S. Yan, X.-H. Xing, X. Sun, B. Jurcik, Progress and perspectives of sludge ozonation as a powerful pretreatment method for minimization of excess sludge production, Water research 43(7) (2009) 1811-1822. https://doi.org/10.1016/j.watres.2009.02.012

[113] N.P. Tanatti, M. Mehmetbaşoğlu, İ.A. Şengil, H. Aksu, E. Emin, Kinetics and thermodynamics of biodiesel wastewater treatment by using ozonation process, Desalination and Water Treatment 161 (2019) 108-115. https://doi.org/10.5004/dwt.2019.24276

[114] K.D. Rauch, A.L. Mackie, B. Middleton, X. Xie, G.A. Gagnon, Biomass recovery method for adenosine triphosphate (ATP) quantification following UV disinfection, Ozone: Science & Engineering 41(2) (2019) 146-155. https://doi.org/10.1080/01919512.2018.1518127

[115] E. GilPavas, I. Dobrosz-Gómez, M.-Á. Gómez-García, Optimization and toxicity assessment of a combined electrocoagulation, H2O2/Fe2+/UV and activated carbon adsorption for textile wastewater treatment, Science of the Total Environment 651 (2019) 551-560. https://doi.org/10.1016/j.scitotenv.2018.09.125

[116] J.A. Malvestiti, E. Fagnani, D. Simão, R.F. Dantas, Optimization of UV/H2O2 and ozone wastewater treatment by the experimental design methodology, Environmental technology 40(15) (2019) 1910-1922. https://doi.org/10.1080/09593330.2018.1432698

[117] L. Rizzo, T. Agovino, S. Nahim-Granados, M. Castro-Alférez, P. Fernández-Ibáñez, M.I. Polo-López, Tertiary treatment of urban wastewater by solar and UV-C driven advanced oxidation with peracetic acid: Effect on contaminants of emerging concern and antibiotic resistance, Water research 149 (2019) 272-281. https://doi.org/10.1016/j.watres.2018.11.031

[118] T.M.H. Nguyen, P. Suwan, T. Koottatep, S.E. Beck, Application of a novel, continuous-feeding ultraviolet light emitting diode (UV-LED) system to disinfect domestic wastewater for discharge or agricultural reuse, Water research 153 (2019) 53-62. https://doi.org/10.1016/j.watres.2019.01.006

[119] L.K. Toke, S.G. Bhadane, T.P. Metkar, A.V. Masali, T.M. Shaikh, Design and Development of Activated Carbon-Based Purification System for Automobile Emissions, Recent Trends in Automation and Automobile Engineering 2(2) (2019).

[120] R. Ligotski, U. Sager, U. Schneiderwind, C. Asbach, F. Schmidt, Prediction of VOC adsorption performance for estimation of service life of activated carbon based filter media for indoor air purification, Building and Environment 149 (2019) 146-156. https://doi.org/10.1016/j.buildenv.2018.12.001

[121] B. Cao, R. Wang, W. Zhang, H. Wu, D. Wang, Carbon-based materials reinforced waste activated sludge electro-dewatering for synchronous fuel treatment, Water research 149 (2019) 533-542. https://doi.org/10.1016/j.watres.2018.10.082

[122] D. Propolsky, E. Romanovskaia, W. Kwapinski, V. Romanovski, Modified activated carbon for deironing of underground water, Environmental Research 182 (2020) 108996. https://doi.org/10.1016/j.envres.2019.108996

[123] S. Periyasamy, I.A. Kumar, N. Viswanathan, Activated Carbon from Different Waste Materials for the Removal of Toxic Metals, Green Materials for Wastewater Treatment, Springer2020, pp. 47-68. https://doi.org/10.1007/978-3-030-17724-9_3

[124] A. Talebi, Y.S. Razali, N. Ismail, M. Rafatullah, H.A. Tajarudin, Selective adsorption and recovery of volatile fatty acids from fermented landfill leachate by activated carbon process, Science of The Total Environment 707 (2020) 134533. https://doi.org/10.1016/j.scitotenv.2019.134533

[125] Z. Heidarinejad, M.H. Dehghani, M. Heidari, G. Javedan, I. Ali, M. Sillanpää, Methods for preparation and activation of activated carbon: a review, Environmental Chemistry Letters (2020) 1-23. https://doi.org/10.1007/s10311-019-00955-0

[126] H.L. Rahman, H. Erdem, M. Sahin, M. Erdem, Iron-Incorporated Activated Carbon Synthesis from Biomass Mixture for Enhanced Arsenic Adsorption, Water, Air, & Soil Pollution 231(1) (2020) 6.

[127] E. Díez, J. Gómez, A. Rodríguez, I. Bernabé, P. Sáez, J. Galán, A new mesoporous activated carbon as potential adsorbent for effective indium removal from aqueous solutions, Microporous and Mesoporous Materials 295 (2020) 109984.

[128] S. Shukla, Rice Husk Derived Adsorbents for Water Purification, Green Materials for Wastewater Treatment, Springer2020, pp. 131-148.

[129] D. Neto, J. Rocha, P.F.L. Vivas, D.P. Singh, R. Freire, Magnetic Nanoparticles in Analytical Chemistry, Magnetochemistry: Materials and Applications 66 (2020) 173. https://doi.org/10.21741/9781644900611-5

[130] M.N. Alves, M. Miró, M.C. Breadmore, M. Macka, Trends in analytical separations of magnetic (nano) particles, TrAC Trends in Analytical Chemistry (2019). https://doi.org/10.1016/j.trac.2019.02.026

[131] A.R. Timerbaev, How well can we characterize human serum transformations of magnetic nanoparticles?, Analyst (2020). https://doi.org/10.1039/C9AN01920K

[132] D. Shi, J. Wallyn, D.-V. Nguyen, F. Perton, D. Felder-Flesch, S. Bégin-Colin, M. Maaloum, M.P. Krafft, Microbubbles decorated with dendronized magnetic nanoparticles for biomedical imaging: effective stabilization via fluorous interactions, Beilstein journal of nanotechnology 10(1) (2019) 2103-2115.

[133] B.T. Thanh, N. Van Sau, H. Ju, M.J. Bashir, H.K. Jun, T.B. Phan, Q.M. Ngo, N.Q. Tran, T.H. Hai, P.H. Van, Immobilization of protein A on monodisperse magnetic nanoparticles for biomedical applications, Journal of Nanomaterials 2019 (2019). https://doi.org/10.1155/2019/2182471

[134] S. Savliwala, A. Chiu-Lam, M. Unni, A. Rivera-Rodriguez, E. Fuller, K. Sen, M. Threadcraft, C. Rinaldi, Magnetic nanoparticles, Nanoparticles for Biomedical Applications, Elsevier2020, pp. 195-221. https://doi.org/10.1016/B978-0-12-816662-8.00013-8

[135] N.V. Vallabani, S. Singh, A.S. Karakoti, Magnetic nanoparticles: current trends and future aspects in diagnostics and nanomedicine, Current drug metabolism 20(6) (2019) 457-472. https://doi.org/10.2174/1389200220666181122124458

[136] Y.P. Yew, K. Shameli, M. Miyake, N.B.B.A. Khairudin, S.E.B. Mohamad, T. Naiki, K.X. Lee, Green biosynthesis of superparamagnetic magnetite Fe3O4 nanoparticles and biomedical applications in targeted anticancer drug delivery system: A review, Arabian Journal of Chemistry 13(1) (2020) 2287-2308. https://doi.org/10.1016/j.arabjc.2018.04.013

[137] N. Irmania, K. Dehvari, G. Gedda, P.J. Tseng, J.Y. Chang, Manganese-doped green tea-derived carbon quantum dots as a targeted dual imaging and photodynamic therapy platform, Journal of Biomedical Materials Research Part B: Applied Biomaterials 108(4) (2020) 1616-1625. https://doi.org/10.1002/jbm.b.34508

[138] K. Dehvari, K.Y. Liu, P.-J. Tseng, G. Gedda, W.M. Girma, J.-Y. Chang, Sonochemical-assisted green synthesis of nitrogen-doped carbon dots from crab shell as targeted nanoprobes for cell imaging, Journal of the Taiwan Institute of Chemical Engineers 95 (2019) 495-503. https://doi.org/10.1016/j.jtice.2018.08.037

[139] D. Song, R. Yang, F. Long, A. Zhu, Applications of magnetic nanoparticles in surface-enhanced Raman scattering (SERS) detection of environmental pollutants, Journal of Environmental Sciences 80 (2019) 14-34. https://doi.org/10.1016/j.jes.2018.07.004

[140] L.M. Sanchez, D.G. Actis, J.S. Gonzalez, P.M. Zélis, V.A. Alvarez, Effect of PAA-coated magnetic nanoparticles on the performance of PVA-based hydrogels developed to be used as environmental remediation devices, Journal of Nanoparticle Research 21(3) (2019) 64. https://doi.org/10.1007/s11051-019-4499-0

[141] M. Sajjadi, M. Nasrollahzadeh, M.R. Tahsili, Catalytic and antimicrobial activities of magnetic nanoparticles supported N-heterocyclic palladium (II) complex: A magnetically recyclable catalyst for the treatment of environmental contaminants in aqueous media, Separation and Purification Technology 227 (2019) 115716. https://doi.org/10.1016/j.seppur.2019.115716

[142] K.A. Abd-Elsalam, M.A. Mohamed, R. Prasad, Magnetic nanostructures: environmental and agricultural applications, Springer2019. https://doi.org/10.1007/978-3-030-16439-3

[143] Y. Shi, D. Jyoti, S.W. Gordon-Wylie, J.B. Weaver, Quantification of magnetic nanoparticles by compensating for multiple environment changes simultaneously, Nanoscale 12(1) (2020) 195-200. https://doi.org/10.1039/C9NR08258A

[144] M. Zulfajri, G. Gedda, C.-J. Chang, Y.-P. Chang, G.G. Huang, Cranberry Beans Derived Carbon Dots as a Potential Fluorescence Sensor for Selective Detection of Fe3+ Ions in Aqueous Solution, ACS omega 4(13) (2019) 15382-15392. https://doi.org/10.1021/acsomega.9b01333

[145] O. Agboola, P. Popoola, R. Sadiku, S.E. Sanni, S.O. Fayomi, O.S. Fatoba, Nanotechnology in Wastewater and the Capacity of Nanotechnology for Sustainability, Environmental Nanotechnology Volume 3, Springer2020, pp. 1-45. https://doi.org/10.1007/978-3-030-26672-1_1

[146] J. Sahu, R.R. Karri, H.M. Zabed, S. Shams, X. Qi, Current perspectives and future prospects of nano-biotechnology in wastewater treatment, Separation & Purification Reviews (2019) 1-20. https://doi.org/10.1080/15422119.2019.1630430

[147] G.N. Hlongwane, P.T. Sekoai, M. Meyyappan, K. Moothi, Simultaneous removal of pollutants from water using nanoparticles: A shift from single pollutant control to multiple pollutant control, Science of the Total Environment 656 (2019) 808-833. https://doi.org/10.1016/j.scitotenv.2018.11.257

[148] W. Guo, Z. Fu, Z. Zhang, H. Wang, S. Liu, W. Feng, X. Zhao, J.P. Giesy, Synthesis of Fe3O4 magnetic nanoparticles coated with cationic surfactants and their applications in Sb (V) removal from water, Science of The Total Environment 710 (2020) 136302. https://doi.org/10.1016/j.scitotenv.2019.136302

[149] L.F.O. Maia, M.S. Santos, T.G. Andrade, R.d.C. Hott, M.C.d.S. Faria, L.C.A. Oliveira, M.C. Pereira, J.L. Rodrigues, Removal of mercury (II) from contaminated water by gold-functionalised Fe3O4 magnetic nanoparticles, Environmental technology 41(8) (2020) 959-970. https://doi.org/10.1080/09593330.2018.1515989

[150] A. Saadat, L. Hajiaghababaei, A. Badiei, M. Ganjali, G. Mohammadi Ziarani, Amino Functionalized Silica Coated Fe3O4 Magnetic Nanoparticles as a Novel Adsorbent for Removal of Pb2+ and Cd2+, Pollution 5(4) (2019) 847-857.

[151] Y. Chen, Z. Zhang, D. Chen, Y. Chen, Q. Gu, H. Liu, Removal of coke powders in coking diesel distillate using recyclable chitosan-grafted Fe3O4 magnetic nanoparticles, Fuel 238 (2019) 345-353. https://doi.org/10.1016/j.fuel.2018.10.125

[152] A. Sajjadi, R. Mohammadi, Fe3O4 magnetic nanoparticles (Fe3O4 MNPs): A magnetically reusable catalyst for synthesis of Benzimidazole compounds, Journal of Medicinal and Chemical Sciences 2(2, pp. 41-75.) (2019) 55-58.

[153] S. Wadhawan, A. Jain, J. Nayyar, S.K. Mehta, Role of nanomaterials as adsorbents in heavy metal ion removal from waste water: A review, Journal of Water Process Engineering 33 (2020) 101038. https://doi.org/10.1016/j.jwpe.2019.101038

[154] L. Huang, R. Shen, R. Liu, Q. Shuai, Thiol-functionalized magnetic covalent organic frameworks by a cutting strategy for efficient removal of Hg2+ from water, Journal of Hazardous Materials 392 (2020) 122320. https://doi.org/10.1016/j.jhazmat.2020.122320

[155] E. Hu, Catalytic ozonation for textile dyeing wastewater treatment and reuse, (2019).

[156] Q.-Y. Wu, Y.-T. Zhou, W. Li, X. Zhang, Y. Du, H.-Y. Hu, Underestimated risk from ozonation of wastewater containing bromide: Both organic byproducts and bromate contributed to the toxicity increase, Water research 162 (2019) 43-52. https://doi.org/10.1016/j.watres.2019.06.054

[157] C. Wolf, A. Pavese, U. von Gunten, T. Kohn, Proxies to monitor the inactivation of viruses by ozone in surface water and wastewater effluent, Water research 166 (2019) 115088.

Materials Research Forum LLC
https://doi.org/10.21741/9781644901144-1

[158] Z. Song, M. Wang, Z. Wang, Y. Wang, R. Li, Y. Zhang, C. Liu, Y. Liu, B. Xu, F. Qi, Insights into heteroatom-doped graphene for catalytic ozonation: Active centers, reactive oxygen species evolution, and catalytic mechanism, Environmental science & technology 53(9) (2019) 5337-5348. https://doi.org/10.1021/acs.est.9b01361

[159] E. Borrelli, The Use of Ozone as Redox Modulator in the Treatment of the Chronic Obstructive Pulmonary Disease (COPD), Oxidative Stress in Lung Diseases, Springer2020, pp. 413-426.

[160] C. Chen, N. Jia, K. Song, X. Zheng, Y. Lan, Y. Li, Sulfur-doped copper-yttrium bimetallic oxides: A novel and efficient ozonation catalyst for the degradation of aniline, Separation and Purification Technology 236 (2020) 116248. https://doi.org/10.1016/j.seppur.2019.116248

[161] M.A. Hassaan, A. El Nemr, Advanced oxidation processes for textile wastewater treatment, International Journal of Photochemistry and Photobiology 2(3) (2017) 85-93.

[162] N. Koutahzadeh, M.R. Esfahani, P.E. Arce, Sequential use of UV/H2O2—(PSF/TiO2/MWCNT) mixed matrix membranes for dye removal in water purification: membrane permeation, fouling, rejection, and decolorization, Environmental Engineering Science 33(6) (2016) 430-440.

[163] A. Sergejevs, C. Clarke, D. Allsopp, J. Marugan, A. Jaroenworaluck, W. Singhapong, P. Manpetch, R. Timmers, C. Casado, C. Bowen, A calibrated UV-LED based light source for water purification and characterisation of photocatalysis, Photochemical & Photobiological Sciences 16(11) (2017) 1690-1699. https://doi.org/10.1039/C7PP00269F

[164] J.-A.V. Sigona, Full contact UV water purification system, Google Patents, 2017.

[165] T. Matsumoto, I. Tatsuno, T. Hasegawa, Instantaneous Water Purification by Deep Ultraviolet Light in Water Waveguide: Escherichia Coli Bacteria Disinfection, Water 11(5) (2019) 968. https://doi.org/10.3390/w11050968

[166] Y. Liao, R.C. Walker, D. Collins, W. Zhang, UV-LED liquid monitoring and treatment apparatus and method, Google Patents, 2019.

[167] E. Magnone, M.-K. Kim, H.J. Lee, J.H. Park, Testing and substantial improvement of TiO2/UV photocatalysts in the degradation of Methylene Blue, Ceramics International 45(3) (2019) 3359-3367. https://doi.org/10.1016/j.ceramint.2018.10.249

Materials Research Forum LLC
https://doi.org/10.21741/9781644901144-1

[168] N. Negishi, M. Sugasawa, Y. Miyazaki, Y. Hirami, S. Koura, Effect of dissolved silica on photocatalytic water purification with a TiO2 ceramic catalyst, Water research 150 (2019) 40-46. https://doi.org/10.1016/j.watres.2018.11.047

[169] X. Li, M. Cai, L. Wang, F. Niu, D. Yang, G. Zhang, Evaluation survey of microbial disinfection methods in UV-LED water treatment systems, Science of the Total Environment 659 (2019) 1415-1427. https://doi.org/10.1016/j.scitotenv.2018.12.344

[170] Y. Liao, R. Walker, D. Collins, D. Theodore, System and method for UV-LED liquid monitoring and disinfection, Google Patents, 2019.

[171] X.-Y. Zou, Y.-L. Lin, B. Xu, T.-Y. Zhang, C.-Y. Hu, T.-C. Cao, W.-H. Chu, Y. Pan, N.-Y. Gao, Enhanced ronidazole degradation by UV-LED/chlorine compared with conventional low-pressure UV/chlorine at neutral and alkaline pH values, Water research 160 (2019) 296-303. https://doi.org/10.1016/j.watres.2019.05.072

[172] N. Dulova, E. Kattel, B. Kaur, M. Trapido, UV-induced Persulfate Oxidation of Organic Micropollutants in Water Matrices, Ozone: Science & Engineering 42(1) (2020) 13-23. https://doi.org/10.1080/01919512.2019.1599711

[173] Y. Liao, R.C. Walker, D. Collins, Device for UV-LED liquid monitoring and treatment, Google Patents, 2020.

[174] M.T. Pickett, L.B. Roberson, J.L. Calabria, T.J. Bullard, G. Turner, D.H. Yeh, Regenerative water purification for space applications: Needs, challenges, and technologies towards' closing the loop', Life Sciences in Space Research 24 (2020) 64-82. https://doi.org/10.1016/j.lssr.2019.10.002

[175] Y. Chen, F. Bai, Z. Li, P. Xie, Z. Wang, X. Feng, Z. Liu, L.-Z. Huang, UV-assisted chlorination of algae-laden water: Cell lysis and disinfection byproducts formation, Chemical Engineering Journal 383 (2020) 123165. https://doi.org/10.1016/j.cej.2019.123165

[176] Y. Guo, H. Lu, F. Zhao, X. Zhou, W. Shi, G. Yu, Biomass-Derived Hybrid Hydrogel Evaporators for Cost-Effective Solar Water Purification, Advanced Materials (2020). https://doi.org/10.1002/adma.201907061

[177] D. Hayashi, J. Yu, P. Dijkstra, M. Van Der Meer, A.J. Hovestad, UV-C water purification device, Google Patents, 2019.

[178] M. Raeiszadeh, F. Taghipour, Microplasma UV lamp as a new source for UV-induced water treatment: Protocols for characterization and kinetic study, Water research 164 (2019) 114959. https://doi.org/10.1016/j.watres.2019.114959

[179] Y. Bilenko, A. Dobrinsky, S. Smetona, M. Shur, R. Gaska, T.J. Bettles, Ultraviolet fluid disinfection system with feedback sensor, Google Patents, 2019.

[180] G.B. Dirisu, U.C. Okonkwo, I.P. Okokpujie, O.S. Fayomi, Comparative analysis of the effectiveness of reverse osmosis and ultraviolet radiation of water treatment, Journal of Ecological Engineering 20(1) (2019) 61-75. https://doi.org/10.12911/22998993/93978

[181] A. Modak, Y.T.C. Osumi, S.F. Pomeranz, A.S. Mendelow, Potable water producing device, Google Patents, 2020.

[182] G. Rijkers, The Crowning with Thorns: How an Intact Skin Forms the First Line of Defense Against Infections, J Vaccines Immunol: JVII-125. DOI 10 (2018). https://doi.org/10.29011/2575-789X.000125

[183] L.A. May, C.M. Woodley, Chemiluminescent method for quantifying DNA abasic lesions in scleractinian coral tissues, Diseases of coral. Wiley-Blackwell, Hoboken (2016) 547-555.

[184] C. P Hoover, Review of Lime-Soda Water Softening. Journal American Water Works Association 29, no. 11 (1937): 1687-1696. https://doi.org/10.1002/j.1551-8833.1937.tb13868.x

[185] H. J. Alexander, M. A. McClanahan. Kinetics of Calcium Carbonate Precipitation in Lime–Soda Ash Softening. *Journal-American Water Works Association* 67, no. 11 (1975): 618-621. https://doi.org/10.1002/j.1551-8833.1975.tb02314.x

[186] J.D. Sheppard, D.G. Thomas, Energy Research and Development Administration (ERDA), 1976. *Water softening process.* U.S. Patent 3,976,569.

[187] K.T. Samarasiri, C.K. Rathnadheera, S. Ishan, P.G. Rathnasiri, Process Optimization for the Lime Soda Water Softening Method for Reducing the Hardness in Well Water (1978).

[188] B. Mustafa, H. W. Walker. Lime-soda softening process modifications for enhanced NOM removal. *Journal of Environmental Engineering* 132, no. 2 (2006): 158-165.

[189] S. B. Applebaum, Characteristic properties of zeolites for water softening. *Journal (American Water Works Association)* 13, no. 2 (1925): 213-220.

[190] S. B. Applebaum, A. S. Behrman. APPLICATIONS OF CARBONACEOUS ZEOLITES TO WATER SOFTENING [with Discussion]. *Journal (American Water Works Association)* 30, no. 6 (1938): 947-978. https://doi.org/10.1002/j.1551-8833.1938.tb12205.x

[191] S. Cinar, B. Beler-Baykal, Ion exchange with natural zeolites: an alternative for water softening?. *Water science and technology* 51, no. 11 (2005): 71-77.

[192] I. Arrigo, P. Catalfamo, L. Cavallari, S. D. Pasquale. Use of zeolitized pumice waste as a water softening agent." *Journal of hazardous materials* 147, no. 1-2 (2007): 513-517. https://doi.org/10.1016/j.jhazmat.2007.01.061

[193] A. R. Loiola, J. C. R. A. Andrade, J. M. Sasaki, L. R. D. Da Silva. Structural analysis of zeolite NaA synthesized by a cost-effective hydrothermal method using kaolin and its use as water softener. *Journal of colloid and interface science* 367, no. 1 (2012): 34-39. https://doi.org/10.1016/j.jcis.2010.11.026

[194] D. A. Bessa, L. S. C. Raquel, C. P. Oliveira, F. Bohn, R. F. do Nascimento, J. M. Sasaki, A. R. Loiola. Kaolin-based magnetic zeolites A and P as water softeners. *Microporous and Mesoporous materials* 245 (2017): 64-72. https://doi.org/10.1016/j.micromeso.2017.03.004

Advances in Wastewater Treatment I
Materials Research Foundations 91 (2021) 37-86

Materials Research Forum LLC
https://doi.org/10.21741/9781644901144-2

Chapter 2

Advanced Oxidation Processes for Wastewater Remediation: Fundamental Concepts to Recent Advances

T.S. Rajaraman[1,3], Vimal Gandhi[2], S.P. Parikh*[1,3]

[1]Chemical Engineering Department, L. D. College of Engineering, Navrangpura, Ahmedabad, Gujarat, India

[2]Department of Chemical Engineering, Dharmsinh Desai University, Nadiad, Gujarat, India

[3]Gujarat Technological University, Chandkheda, Ahmedabad, Gujarat, India

Abstract

Industrialization and modernization in recent times have led to a water crisis across the world. Conventional methods of water treatment like physical, chemical and biological methods which comprise of many commonly used techniques like membrane separation, adsorption, chemical treatment etc. have been in use for many decades. However, problems like sludge disposal, high operating costs etc. have led to increased focus on Advanced Oxidation Processes (AOPs) as alternative treatment methods. AOPs basically involve reactions relying on the high oxidation potential of the hydroxyl (OH$^•$) free radical. They have the potential to efficiently treat various toxic, organic pollutants and complete degradation of contaminants (mineralization) of emerging concern. Many different types of homogenous as well as heterogenous AOPs have been studied viz: UV/H$_2$O$_2$, Fenton, Photo-Fenton, Sonolysis, Photocatalysis etc. for treatment of a wide variety of organic pollutants. Different AOPs are suitable for different types of wastewater and hence proper selection of the right technique for a particular type of pollutant is required. The inherent advantages offered by AOPs like elimination of sludge disposal problems, operability under mild conditions, ability to harness sunlight, non selective nature (ability to degrade all organic and microbial contamination) etc. have made it one of the most actively researched areas in recent times for wastewater treatment. Despite the benefits and intense research, commercial applicability of AOPs as a practical technique for treating wastewater on a large scale is still far from satisfactory. Nevertheless, positive results in lab scale and pilot plant studies make them a promising water treatment technique for the future. In the present chapter, an attempt has been made to discuss all aspects of AOPs beginning with the fundamental concepts, classification, underlying mechanism, comparison, commercialization to the latest developments in AOPs.

Keywords

Advanced Oxidation Processes (AOPs), Wastewater Treatment, Degradation, Organic Pollutants, Hydroxyl Radicals, Chemical Oxygen Demand (COD)

Contents

Materials Research Forum LLC
https://doi.org/10.21741/9781644901144-2

1.   Introduction

Wastewater treatment has been the focus of many researchers and scientists for many decades now. Rapid industrialization and population explosion have made the conservation and treatment of water a high priority activity throughout the world. Wastewater can be broadly classified as industrial, municipal and agricultural and it mainly consists of 99.9% water by weight and 0.1 % dissolved/suspended impurities [1]. Wastewater treatment has been around for over a century. The earlier methods of treatment, in the 1900's, usually involved spreading of sewage water in large areas on land where it decayed under the action of micro-organisms. Subsequent degradation of land forced us to set higher targets for wastewater treatment in the later years like removal of suspended impurities, treatment of biodegradable and microbial pollutants etc [2]. Nowadays, wastewater treatment technologies can be mainly classified as physical, chemical, and biological processes (Fig. 1) [1].

As seen in Fig. 1, many methods have been employed over the last few decades for the treatment of wastewater. Some of these methods include chemical treatment methods, adsorption, membrane separation etc. However, all these conventional treatment technologies have many drawbacks which need to be overcome. For example, chemical methods involve formation of large amounts of sludge and require large quantities of chemical. Membrane technologies have operational difficulties and involve high capital costs. Similarly, adsorption techniques also involve only the transfer of pollutant from one medium to another and again have sludge disposal problems [3]. These problems have driven the research towards developing an alternative treatment strategy where the Advanced Oxidation Processes (AOPs) have taken the centre stage. The AOPs have revolutionized the water treatment field as the process involves direct elimination of the pollutants (complete mineralization) rather than their transfer from one medium to another.

Materials Research Forum LLC

https://doi.org/10.21741/9781644901144-2

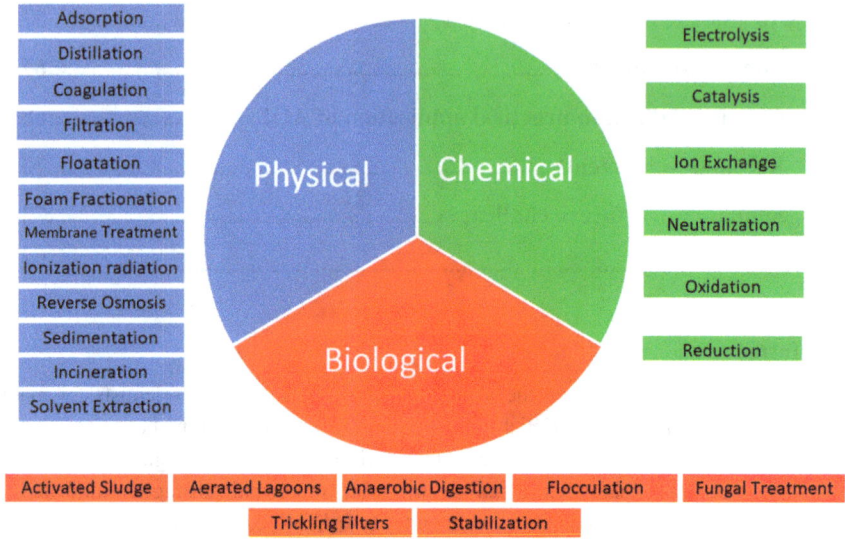

Figure 1. Flowchart depicting the various physical, chemical and biological processes.

AOPs are methods which utilize the high oxidizing potential of the hydroxyl free radicals (OH$^{\bullet}$) (also sulphate radicals in some cases) in degradation of various types of pollutants in wastewater. Although other species are also reported, hydroxyl radicals are the main species which are responsible for the degradation of the various contaminants [4]. AOPs have been around since the 1980s where these were first used for potable water after which their application was then extended to various types of wastewaters [5]. As the years went by, AOPs were applied to different types of wastewater and effluents mainly for the reduction of their organic content. The non specific nature of these radicals means that the process could be used for the degradation of a host of organic pollutants thereby proving their versatility.

The following sections broadly cover the different types of AOPs with their mechanism, advantages and limitations. Comparison of different AOPs and their scope for commercialization has also been discussed in the later sections. At the end, recent developments in this field and future challenges for AOPs as an alternative technology for wastewater treatment has also been included.

2. Mechanism and classification

2.1 General mechanism

Although different AOPs exhibit slightly different mechanisms on how they break down the target pollutant, the general steps which are followed remain the same. The basic steps involved in any AOP are as follows [1]:

1) Formation of highly reactive free radicals (oxidants) like $OH^•$, superoxide radical, sulphate radical etc.
2) Reaction between these free radicals and the organic contaminants of the wastewater to degrade them into simpler compounds.
3) Further oxidation of these intermediate products (mineralization) into simpler compounds like carbon dioxide, water, inorganic salts, etc.

A schematic representation of these steps is shown in Fig. 2. It is important to note that the extent of these 3 steps varies between the AOPs and is highly dependent on various factors like target pollutant, reaction conditions, free radical concentration etc. The hydroxyl radical ($OH^•$) is highly reactive and non selective in reacting with organic compounds. It reacts by hydrogen abstraction, by addition to unsaturated bonds and aromatic rings, or by electron transfer. The common pathway by hydrogen abstraction is as follows (Eqs. 1-4) [6] :

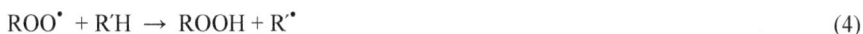

$$OH^• + RH \rightarrow H_2O + R^• \tag{1}$$

$$R^• + H_2O_2 \rightarrow ROH + OH^• \tag{2}$$

$$R^• + O_2 \rightarrow ROO^• \tag{3}$$

$$ROO^• + R'H \rightarrow ROOH + R'^• \tag{4}$$

where, RH and R'H are the organic molecules

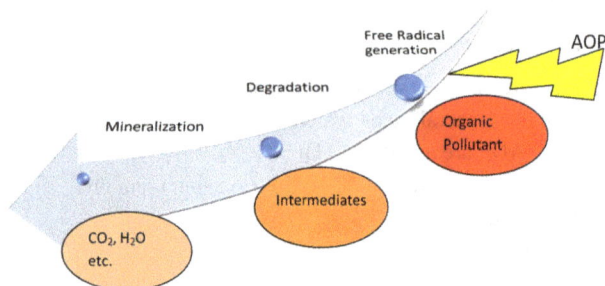

Figure 2. Schematic diagram representing generalized mechanism of AOPs.

In all AOPs, apart from the free radical mechanism, other mechanisms may also be present for the degradation of the target component. And this maybe a dominant or insignificant mechanism depending on the AOP [5].

Although hydroxyl radicals are the most commonly reported free radical in the literature, sulphate radicals based AOPs are also gathering attention in recent years. The associated mechanism, however, has been covered in a separate section.

2.2   Classification

Classification of AOPs have been proposed in different ways in the literature. Different types of classifications include those based on the type (ozone based, UV based, electrochemical AOPs, catalytic AOPs etc.) [7], based on radicals involved (hydroxyl radicals based or sulphate radicals based) [5], based on whether AOP is homogenous or heterogeneous [8], photochemical or non photochemical [9] etc.

Classification of AOPs becomes a little difficult at times when the same AOP can come under multiple categories. An AOP like O_3/UV (ozonation under UV light) can be classified as a UV based as well as ozone based AOP. Fig. 3 depicts a classification which takes into account more than one type of criteria mentioned above. The base criteria have been taken as whether the AOP is homogenous or heterogeneous. However, AOPs like Wet Air Oxidation, Microwave (MW) and electro-Fenton have been placed in the grey area as both homogenous and heterogeneous variants have been explored in the literature in separate studies. It must be mentioned that even this classification will find the need to be modified in the future as more and more different combination of AOPs are being explored everyday for synergistic effects which would make it difficult to place them under one roof.

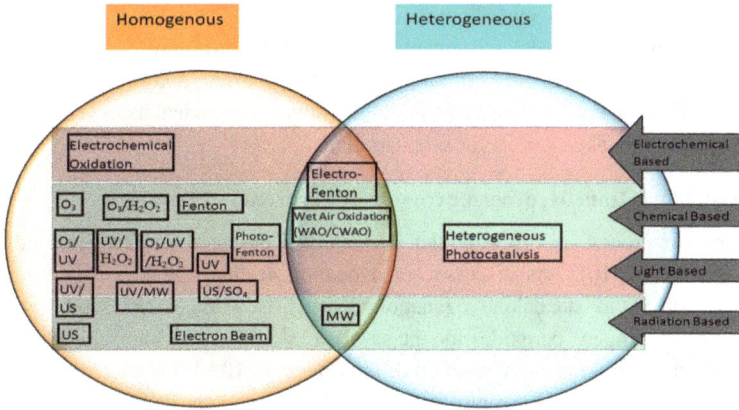

Figure 3: Classification of AOPs (US: Ultrasound, MW: Microwave, SO₄: sulphate radicals based AOPs, WAO: Wet Air Oxidation, CWAO: Catalytic Wet Air Oxidation).

3. Various advanced oxidation processes

3.1 Ozonation (O_3)

Ozone is a highly reactive gas which is partially soluble in water. The ozonation process for the treatment of wastewater involves two mechanisms viz: direct and indirect. In the direct route, the ozone molecule itself reacts directly with the organic molecules present in the wastewater thereby leading to its degradation (owing to its high standard redox potential (E_0) of 2.07 V) [10]. The second route is indirect in the sense that the ozone first gets decomposed to produce hydroxyl radicals which then attack the organic molecules and degrade them. The second route is the preferred route as the reaction rate in the hydroxyl radical mediated reactions is several orders of magnitude higher than the rates obtained in the reactions with molecular ozone (direct route) [11]. The main reactions by which the ozone decomposes in the indirect route are as follows [10]:

Acidic to neutral pH:

$$O_3 + OH^- \rightarrow HO_2 + O_2^{\bullet -} \qquad k_{i1} = 70 \ M/s \qquad (5)$$

Alkaline pH:

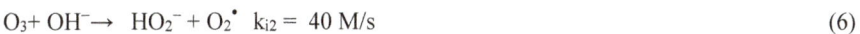

$$O_3 + OH^- \rightarrow HO_2^- + O_2^{\bullet} \quad k_{i2} = 40 \ M/s \qquad\qquad\qquad (6)$$

$$O_3 + HO_2^- \rightarrow HO_2^{\bullet} + O_3^{\bullet -} \qquad k_{i3} = 2.2 * 10^6 \ M/s \qquad (7)$$

The ozonation process has some limitations. The production of hydroxyl radicals for the indirect route is much lower as compared to other AOPs. Moreover, ozone by itself is selective where it reacts preferentially with only ionized and dissociated organic compounds through the direct route [5]. Additionally, toxicological aspects of ozone as a reagent needs to be taken care of for safe handling.

3.2. O_3/H_2O_2 (Ozone/Hydrogen Peroxide or Peroxone)

The peroxone process is probably the best studied and best implemented advanced oxidation process [10]. As discussed in ozonation section, the low concentration of hydroxyl radicals in standalone ozonation process is a major drawback. However, addition of hydrogen peroxide to the system significantly increases the radical concentration. Even a low concentration of H_2O_2 (10^{-5} to 10^{-4} M) is enough to accelerate the decomposition of ozone in aqueous medium due to which more hydroxyl radicals are generated [10]. The reactions involved are as follows (Eqs. 8 and 9) [5]:

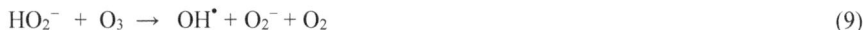

$$H_2O_2 \rightarrow HO_2^- + H^+ \tag{8}$$

$$HO_2^- + O_3 \rightarrow OH^\bullet + O_2^- + O_2 \tag{9}$$

The O_3/H_2O_2 system overcomes the inherent drawbacks of the standalone ozonation process like reducing the bromate formation, increased efficiencies and shorter reaction times for degradation of organic substances [12].

Depending on the target pollutant, optimal dosages of ozone and hydrogen peroxide can be found for best results. However, residual peroxide has been reported to cause some problems in the proper assessment of contaminant removal [13]. Moreover, hydrogen peroxide also acts as radical scavenger. So, excess peroxide may be counterproductive as it might reduce the hydroxyl radical concentration.

3.3.   UV

UV photolysis is one of the most commonly used technique for water treatment. In the direct photolysis approach, the contaminant absorbs the UV radiation which subsequently leads to its degradation. However, the low photo-dissociation efficiencies of the direct photolysis by UV light often limits its practical use [14]. As with the standalone ozonation, the standalone UV photolysis involves only the direct degradation of the contaminants without the involvement of the hydroxyl free radicals. Hence, hydroxyl radical based indirect approach is the preferred one where in situ generation of hydroxyl

Materials Research Forum LLC
https://doi.org/10.21741/9781644901144-2

radicals by the photolysis of ozone or hydrogen peroxide results in more efficient degradation of the contaminants.

However, some target pollutants are more susceptible to the direct photolysis and in those cases direct photolysis might be the preferred method where degradation can take place without the addition of external reagents (O_3/H_2O_2). In some cases, the pollutant might be equally degraded by both direct and indirect routes and in such cases optimal conditions for best efficiencies might be needed [14].

Molecules of different species have differing peak absorption wavelength and hence the efficiency of any AOP involving UV radiation is quite dependent on the target pollutant and the wavelength of irradiation.

3.4 O_3/UV

The standalone ozonation, as seen earlier, has some limitations such as slow removal of compounds like ammonia, high cost of ozone, toxicity of some intermediate products formed (like bromate) etc [12]. In an effort to increase the efficiency of the ozonation process, UV radiation has been used in conjunction with ozone for wastewater treatment. The O_3/UV process can lead to the degradation of molecules by 3 ways: 1) ozonation 2) photolysis of ozone by UV which generates hydroxyl radicals 3) and photolysis of the pollutant molecule itself [3]. The photolysis of dissolved ozone takes place as per the Eq. 10 [10].

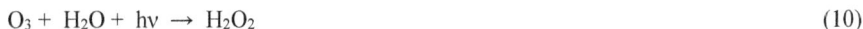

$$O_3 + H_2O + h\nu \rightarrow H_2O_2 \tag{10}$$

This hydrogen peroxide formed can then lead to generation of hydroxyl radicals by ozone decomposition as per reaction (8) and (9), or can undergo further photolysis to form more hydroxyl radicals by itself.

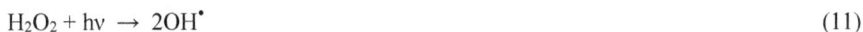

$$H_2O_2 + h\nu \rightarrow 2OH^\bullet \tag{11}$$

O_3/UV process has been widely applied for the treatment of wastewater containing a variety of pollutants.

3.5. UV/H_2O_2

The ultraviolet radiation in the UV/H_2O_2 system serves a dual role in the water disinfection process where the UV radiation can aid in the photolysis process of the peroxide for generating the hydroxyl radicals (Eq. 12) as well as physically inactivating the microorganisms by itself [15].

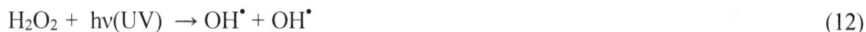

$$H_2O_2 + h\nu(UV) \rightarrow OH^\bullet + OH^\bullet \tag{12}$$

An optimal concentration of H_2O_2 is desired as too high a concentration of H_2O_2 leads to scavenging of the generated free radicals by the peroxide molecules themselves through the below mentioned reactions (Eq. 13-15)).

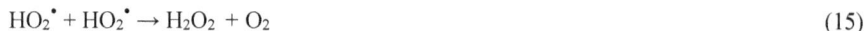

$$OH^\bullet + H_2O_2 \rightarrow HO_2^\bullet + H_2O \tag{13}$$

$$HO_2^\bullet + H_2O_2 \rightarrow OH^\bullet + H_2O + O_2 \tag{14}$$

$$HO_2^\bullet + HO_2^\bullet \rightarrow H_2O_2 + O_2 \tag{15}$$

In general, the degradation of organic compounds by UV/H_2O_2 occurs in two ways: direct photolysis and oxidative attack by hydroxyl radicals [16].

The hydroxyl radical attack follows a second order reaction. However, a pseudo first order reaction can be assumed by considering a constant concentration of hydroxyl radicals over the range of reactions. Various factors like alkalinity, pH, other organic and inorganic matter in the water matrix affect the rate of degradation.

3.6 $UV/O_3/H_2O_2$

Addition of hydrogen peroxide to the O_3/UV system is cost effective and leads to further increase in the generation of hydroxyl radicals.

Many studies report that the combination of ozone, hydrogen peroxide and UV light indeed shows the best results in terms of degradation/mineralization of contaminants. Peternel et al. [17] have compared the decolorization/mineralization efficiencies of different processes like UV, UV/H_2O_2, UV/O_3 and $UV/O_3/H_2O_2$ for the removal of RR45 commercial dye. The results revealed that except for UV process, all the other systems were able to remove the color completely. Moreover, the $UV/O_3/H_2O_2$ system showed the best mineralization performance as compared to other systems. For a 1 hr partial mineralization, the systems with increasing order of extent of mineralization was UV < UV/H_2O_2 < UV/O_3 < $UV/O_3/H_2O_2$. The $UV/O_3/H_2O_2$ was reported to be 4 times faster for complete mineralization as compared to UV/H_2O_2 process. Similar studies have also been carried out by Kusic et al. [18] where phenol degradation capabilities of different systems were compared (O_3,O_3/H_2O_2, UV/H_2O_2, UV/O_3 and UV/O_3/H_2O_2). Again, the results suggested that the UV/O_3/H_2O_2 system was the most effective one with 58% TOC removal in 1 hr as compared to 44.3% removal in the next best UV/O_3 system. Cost effectiveness of different systems was also compared and interestingly, the UV/O_3/H_2O_2 was found to be the most cost effective one.

This clearly emphasizes the fact that combination of ozone, hydrogen peroxide and UV light not only results in enhanced degradation of organics but also are the most economical ones.

3.7 Fenton's process

Fenton's reagent is a combination of Ferrous salt and H_2O_2 (discovered by H.J.H. Fenton more than 100 years ago). It was quickly realized that the Fenton's reagent is a strong oxidant for a number of organic substances [19]. It was later found that the hydroxyl radicals generated as per the Eq. (16)-(21) [20] are responsible for the strong oxidation reactions.

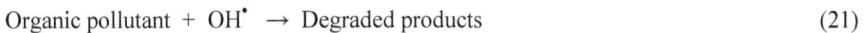

$$Fe^{2+} + H_2O_2 \rightarrow Fe^{3+} + OH^{\bullet} + OH^{-} \tag{16}$$

$$OH^{\bullet} + H_2O_2 \rightarrow HO_2^{\bullet} + H_2O \tag{17}$$

$$Fe^{2+} + OH^{\bullet} \rightarrow Fe^{3+} + OH^{-} \tag{18}$$

$$Fe^{3+} + HO_2^{\bullet} \rightarrow Fe^{2+} + O_2 + H^{+} \tag{19}$$

$$OH^{\bullet} + OH^{\bullet} \rightarrow H_2O_2 \tag{20}$$

$$\text{Organic pollutant} + OH^{\bullet} \rightarrow \text{Degraded products} \tag{21}$$

The reaction between hydrogen peroxide and Fe^{2+} requires acidic conditions and hence sometimes the reaction is shown as follows (Eq. 22) [21]:

$$Fe^{2+} + H_2O_2 + H^{+} \rightarrow Fe^{3+} + OH^{\bullet} + OH^{-} \tag{22}$$

Many studies have discussed the degradation ability of the Fenton's reagent. Sun et al. [22] have studied the degradation of p-nitroaniline (PNA) by Fenton oxidation where they found that the rate of reaction was strongly affected by parameters such as pH, H_2O_2 and Fe^{2+} dosage, temperature and initial PNA concentration.

The Fenton's reaction has certain shortcomings viz: 1) Iron sludge generation 2) narrow working range of pH 3) high cost and risk associated with storage, transportation and handling of various reagents used [21–23]. The iron sludge generation is a major problem as it acts as a major barrier for commercialization of this Fenton Process. The Photo-Fenton process, in some ways, overcomes this barrier.

3.8   Photo-Fenton process

The efficiency of the Fenton's process was found to increase by the UV/light irradiation. This then came to be known as the Photo-Fenton process. In the Photo-Fenton reaction, in addition to the reaction taking place in the Fenton process, an additional reaction (Eq. (23)) takes place in the presence of light which enhanced the catalytic capability of the Fenton process [24]. The hydrogen peroxide also yields additional hydroxyl radicals (Eq. (24)) for subsequent degradation reactions and the sludge formation is minimized as the $Fe(OH)_2$ is converted back into Fe^{2+} ions which can again participate in the production of free radicals through Fenton reaction. This synergistic effect of UV light and Fenton process has drawn great attention owing to the reduction of sludge formation.

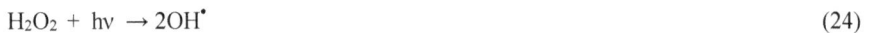

$$Fe(OH)_2 + h\nu \ \rightarrow \ Fe^{2+} \ + \ OH^\bullet \tag{23}$$

$$H_2O_2 \ + \ h\nu \ \rightarrow 2OH^\bullet \tag{24}$$

The photo-Fenton process has the potential to use sunlight as the light source in place of UV light. Since sunlight only contains a small percentage of UV region in its spectrum, the degradation efficiencies will obviously be lower. However, high operating costs associated with UV light can be saved.

3.9   Heterogeneous Photocatalysis

Photocatalysis is a chemical reaction induced in a photocatalyst upon photoabsorption, which remains unchanged during the reaction. Principally, it is a "green" technology where the photocatalyst, upon irradiation, creates electron–hole pairs, which generate free radicals (e.g. hydroxyl radicals: $OH^\bullet$) to be used in secondary reactions. It caters to a wide range of applications, with wastewater treatment being one of the main areas.

Photocatalysis utilizes light source (artificial or solar) and oxygen from air and takes place at ambient temperature and pressure [25].

The photocatalyst, usually solid, is what makes this process heterogeneous in nature. Heterogeneous photocatalysis broadly consists of three main steps [26]:

1) light absorption

2) electron hole separation and

3) surface reaction

Light with energy greater than bandgap of photocatalyst (like TiO_2) excites electrons from Valence Band (VB) to Conduction Band (CB) thereby leaving positive holes behind in VB as per Eq. (25). These photogenerated electrons and holes which have been separated can then take part in oxidation and reduction reactions. Electrons can react with oxygen to give superoxide radicals (O^{2}-*) whereas the holes can react with water to yield hydroxyl radicals ($OH^•$)(Eq. 27 and Eq. 28) [27]. These free radicals can then react with a host of other species through taking part in redox reactions for organic pollutant degradation (Eq. 29 and Eq. 30). However, the photogenerated electrons and holes can recombine with each other as per Eq. (26). Fig. 4 represents various steps involved in heterogenous photocatalysis.

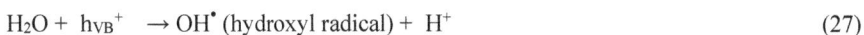

$$\text{Photocatalyst} + h\nu \rightarrow h_{VB}^+ + e_{CB}^- \tag{25}$$

$$h_{VB}^+ + e_{CB}^- \rightarrow \text{energy (heat)} \tag{26}$$

$$H_2O + h_{VB}^+ \rightarrow OH^• \text{ (hydroxyl radical)} + H^+ \tag{27}$$

$$O_2 + e_{CB}^- \rightarrow O^{2-•} \text{ (superoxide radical)} \tag{28}$$

$$OH^• + \text{pollutant} \rightarrow \text{Intermediates} \rightarrow H_2O + CO_2 \tag{29}$$

$$O^{2-•} + \text{pollutant} \rightarrow \text{Intermediates} \rightarrow H_2O + CO_2 \tag{30}$$

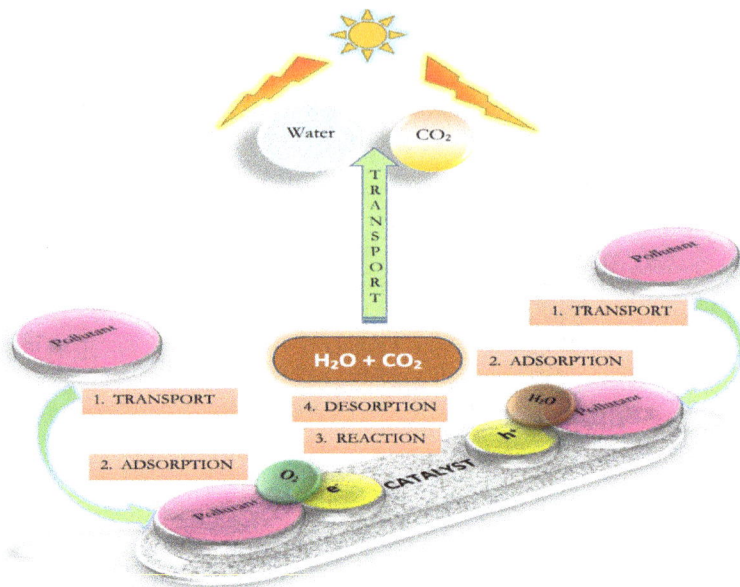

Figure 4: Schematic representation of steps in Heterogenous Photocatalysis (Reprinted with permission from Ref. [27]. Copyright 2017, Elsevier).

A positive hole in the titanium dioxide , TiO_2^+ has the highest oxidation power of all species (3.19 eV, higher than the hydroxyl radical and the fluoride radicals) and hence holes have the ability to oxidize most of the chemicals [9].

Heterogenous photocatalysis is one of the most researched areas in the scientific community. Conventional photocatalysts like titanium dioxide (TiO_2) are active only in the UV region of the spectrum. However, tremendous research with a motive to develop photocatalysts which can operate under sunlight have led to novel visible light active photocatalysts. Some of the visible light photocatalysts along with their performances in pollutant degradation are summarized in Table 1.

Some other notable visible light active photocatalysts include WO_3, $BiVO_4$, Ag_2ZnGeO_4, Cr_2O_3/SnO_2, Graphene- CdS, Pt/CdS–Titania Nanotube Arrays etc. [27].

Table 1: Some TiO₂ based recyclable visible light photocatalysts and their performance.

Photocatalyst	Light Source	Target Pollutant	Degradation	Ref.
TiO_2–Cu_2O	Sunlight	Sodium Dodecyl-Benzenesulfonate (SDBS)	100% removal of SDBS from the solution in 1 hr.	[28]
TiO_2-PANI /Cork composite (polyaniline)	Natural Sunlight (10 am to 2 pm)	Methyl Orange, Phenol, Nitrophenol, Toluendene, Salicylic acid and Benzoic acid	TiO_2–PANI (50 wt.%)/Cork was optimum where 95.2% removed in 210 mins of sunlight radiation.	[29]
$Ag/TiO_2/Fe_3O_4$	Simulated visible light (30W, Philips, the Netherlands)	Methyl Orange (MO)	MO degradation of magnetic $Ag/TiO_2/Fe_3O_4$ was 60.06% after 120 min, as compared to that of TiO_2 (25.94%) and Ag/TiO_2 (38.52%).	[30]
$FeWO_4$ nanorods	UV–Visible light	Methyl Orange MO	96.8 % in 120 minutes	[31]

3.10   Ultrasound AOP (Sonolysis)

Sonolysis is an advanced oxidation process that consists of generation of hydroxyl radicals through ultrasound for degradation of contaminants. The sonication process has many inherent advantages including elimination of reagent requirement, ability for selective degradation, ease of use, short contact times etc [32]. Basically, the ultrasound (20- 10,000 kHz) can be classified into three types viz: low frequency (20-100kHz), high frequency (200-1000kHz) and very high frequency (5000-10,000KHz) [33].

In sonolysis or sonochemical process, ultrasonic sound waves create 'acoustic cavitation' when they are transmitted through an aqueous solution. In acoustic cavitation, formation, growth and collapse of bubbles takes place. The bubbles are generated from the compressive and expansion cycles of the ultrasound wave. These micro-sized bubbles collapse in split seconds which act as a localized 'hot spot' microreactors where huge magnitude of energy is released due to extremely high temperature and pressure (5000K and 1000 atm) [33].

Materials Research Forum LLC
https://doi.org/10.21741/9781644901144-2

The relevant reactions are as follows:

$$H_2O +))) \rightarrow OH^{\bullet} + OH^- \tag{31}$$

$$H_2O +))) \rightarrow \tfrac{1}{2}H_2 + \tfrac{1}{2}H_2O_2 \tag{32}$$

where, ')))))' represents ultrasound.

Many sonolysis studies are available in the literature for the degradation of organic as well as microbial contaminants. In recent years, many studies have been directed towards using sonolysis in combination with other AOPs. One of such commonly found hybrid AOP is the sonophotocatalysis (also known as sonophotochemical process) where a semiconductor photocatalyst is irradiated with UV irradiation in the presence of ultrasound waves.

The relevant reactions are as follows (Eq. 33-40) [34]:

Water sonolysis:

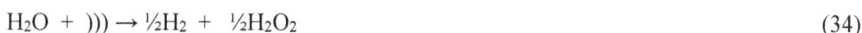

$$H_2O +))) \rightarrow OH^{\bullet} + OH^- \tag{33}$$

$$H_2O +))) \rightarrow \tfrac{1}{2}H_2 + \tfrac{1}{2}H_2O_2 \tag{34}$$

Semiconductor photocatalysis (TiO$_2$ semiconductor)

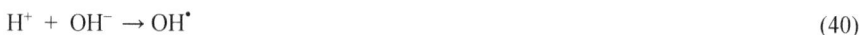

$$TiO_2 + h\nu \rightarrow TiO_2^- + OH^{\bullet} \ (or\ TiO_2^+) \tag{35}$$

$$TiO_2^- + O_2 + H^+ \rightarrow TiO_2 + HO_2 \tag{36}$$

$$TiO_2^- + H_2O_2 + H^+ \rightarrow TiO_2 + H_2O + OH^{\bullet} \tag{37}$$

$$TiO_2^- + 2H^+ \rightarrow TiO_2 + {}^{\bullet}H_2 \tag{38}$$

$$H^+ + H_2O \rightarrow OH^{\bullet} + H^+ \tag{39}$$

$$H^+ + OH^- \rightarrow OH^{\bullet} \tag{40}$$

Here the hybrid of photocatalysis and sonolysis leads to enhanced production of the free radicals thereby leading to higher efficiencies in the degradation reactions.

Similarly, many other combinations have also been reported like the sonolysis/hydrogen peroxide (US/H_2O_2), sonolysis/Fenton (US/Fe^{2+}) or sono-Fenton for the degradation of Coomassie Brilliant Blue (CBB) [35].

3.11 Microwave AOPs

Microwave based AOPs involve the application of highly energetic radiation in the microwave range (300 MHz - 300 GHz) for degradation of organic components present in the water [36].

Both homogenous and heterogeneous based microwave based AOPs exist. In heterogeneous processes, the microwave induces formation of hotspots on the edges and active sites of solid substrates including solid catalysts. Hydroxyl and superoxide anion radicals are formed due to these hotspots which subsequently lead to the degradation of pollutant molecules. In case of homogeneous processes, the same free radicals are produced by the microwave induced activation of chemical oxidants [36].

The response of any compound to microwave is based on the dipole interactions and/or ionic conduction. Microwave energy in itself is not sufficient to induce chemical reactions and hence microwave radiation is often involved in Microwave Assisted Processes (MAP) where it increases the rate of reactions of other AOPs when combined with them ($MW/UV/TiO_2$, MW/UV, $MW/UV/H_2O_2$, MW/photocatalysis etc.) [37]. Many studies have been carried out in the last two decades in microwave assisted processes for wastewater treatment. A review of many such studies is available in literature [38]. The importance of microwave is particularly emphasized in studies carried out by Jou et al. [39] where degradation of pentachlorophenol (PCP) was investigated. Here, the degradation was carried out in the presence of zero valence iron (Fe^0) and microwave radiation. On comparison, the microwave radiation having 700W energy along with (Fe^0) could remove 99.9% of PCP (1000 ppm solution) whereas only 3% removal efficiency was seen without MW. Studies by Parolin et al. [40] give a comparison of various MW assisted and other standalone processes. Their result suggest that for the degradation of the azo dye tartrazine, $UV/H_2O_2/MW$ process led to 8 times enhancement in the removal efficiency as compared to other processes (MW, H_2O_2, H_2O_2/MW and UV/MW).

Despite promising results, unfortunately, most of the applied microwave energy is converted into heat. Beside the low electrical efficiency, cooling devices are required to prevent overheating of treated water [7].

3.12 Supercritical Water Oxidation (SCWO)

Supercritical water oxidation (SCWO) is an advanced oxidation process which makes use of the desirable properties of supercritical condition for the oxidation of organics in wastewater. In this process, the oxidation reactions take place above the critical point of water (647.3 K and 22.12 MPa). As the polarity of water changes at the supercritical condition, it is completely miscible with the organics and oxygen creating a homogenous reaction medium which is highly desirable for the oxidation reactions [41]. Under appropriate conditions (temperatures, pressures and residence times), SCWO has the ability to completely destroy any pollutant in residence time of less than 1 minute. In general, for waters having organic content < 1%, other AOPs are suited whereas for highly concentrated wastewaters with organic content between 1 -20%, SCWO is the better option [42].

Many studies on SCWO as an AOP exist in the literature [42,43]. Despite all the inherent benefits, SCWO still has many disadvantages in terms of operational problems. Problems of corrosion, deposition, high cost etc. have been responsible for the fact that very few studies have been extended beyond laboratory level research. However, Vadillo et al. [44] have analyzed all these limitations and have reported that slight operational changes can make SCWO a feasible AOP for treatment of real industrial wastewater.

3.13 Gamma-ray, X-ray and Electron Beam Based Processes

Gamma-ray, X-ray and Electron Beam Based Processes come under the radiation based AOPs (electron beam radiation) where radiolysis of water is the main source of generation of free radicals. The products of radiolysis of water react with the organic pollutants [45]. Radiolysis of water mainly involves two reactions (Eq. 41 and 42):

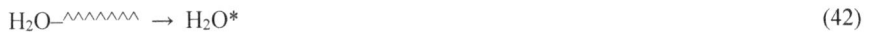

$$H_2O\text{--}\wedge\wedge\wedge\wedge\wedge\wedge \rightarrow H_2O^{+\bullet} + e^- \tag{41}$$

$$H_2O\text{--}\wedge\wedge\wedge\wedge\wedge \rightarrow H_2O* \tag{42}$$

The water radical cation ($H_2O^{+\bullet}$) and electronically excited water molecules (H_2O*) may be involved in spur reactions [45]. The penetration depth of the electrons thus generated depend on the energy of the incident radiation due to which water is irradiated in the form of a thin film for high oxidizing power [7]. Many studies have indicated the potential of these electron beam process to be used as AOP. Studies by Vahdat et al. [46] have confirmed that electron beam radiation is effective for decolorization and mineralization of reactive dyes. In similar lines, many such studies have been carried out where disinfection and radiolytic decomposition of organics have been reported. Also, the technique had been used in combination with other AOPs as well (like ozone/electron beam process) to make it more cost effective [47].

3.14 Wet Air Oxidation

Wet Air Oxidation (WAO) is a type of advanced oxidation process which uses elevated conditions of temperature (125-320 °C) and pressure (0.5 -20 MPa) for the oxidation reactions whereby, toxic, non biodegradable pollutants in wastewater are completely mineralized to carbon dioxide and water or are decomposed to simpler biodegradable products [48]. Gaseous oxygen is used as an oxidant. WAO is well suited for wastewaters containing high organic loading (10-100 g/L). Application of appropriate catalysts to WAO, i.e., Catalytic Wet Air Oxidation (CWAO) further improves the degradation ability of the process along with making it more cost effective. Both homogenous and heterogenous CWAO have been studied. However, in homogenous CWAO, additional step is required for the recovery of catalysts (like dissolved copper ions) and hence heterogenous CWAO has gathered much attention in the last few years as the extra separation step is eliminated [49]. Various noble metals (Ru, Pd, Ce, Pt etc.), metal oxides (CeO_2, SiO_2, TiO_2 etc) and mixed oxides have been used as catalysts over the years. An extensive review of such studies for treating phenolics, carboxylic acids and nitrogen containing compounds is available in the literature [49,50].

WAO becomes self sustaining when feed COD greater than 20000 as high pressure steam produced in the process itself acts as a source of energy [48].

3.15 Electrochemical Oxidation

Electrochemical oxidation based AOPs have gained popularity among the scientific community. These are basically electrochemical processes which use oxidation reactions to degrade the pollutants present in the water [51]. Anodic oxidation was one of the first studied process which involved direct electrolysis of pollutants. Later on, other processes like anodic oxidation with electrogenerated H_2O_2, electro-Fenton process, photo-electro-Fenton processes etc. have been widely studied.

3.15.1 Anodic oxidation (AO)

AO involves the oxidation of pollutants by: (i) direct electron transfer to the anode surface M, (ii) heterogeneous free radicals produced as intermediates of oxidation of water to oxygen, such as physisorbed $OH^{\bullet}$ at the anode surface $M(OH^{\bullet})$ (Eq. 43), oxidants like H_2O_2 produced from $M(OH^{\bullet})$ dimerization and O_3 formed from water discharge at the anode surface by Eq. 44 and Eq. 45 respectively, and/or (iii) other weaker oxidant agents electrochemically produced from ions existing in the bulk [52].

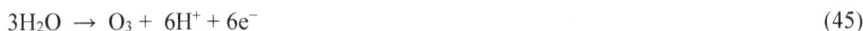

$$M + H_2O \rightarrow M(OH^\bullet) + H^+ + e^- \tag{43}$$

$$2M(OH^\bullet) \rightarrow 2MO + H_2O_2 \tag{44}$$

$$3H_2O \rightarrow O_3 + 6H^+ + 6e^- \tag{45}$$

Anodic oxidation with electrogeneration of H_2O_2 or electroperoxidation involves generation of H_2O_2 at the cathode surface by reduction of oxygen in acidic/neutral media as per Eq. 46.

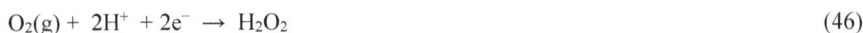

$$O_2(g) + 2H^+ + 2e^- \rightarrow H_2O_2 \tag{46}$$

H_2O_2 is a moderately strong oxidant. For high efficiency for electrogenerating H_2O_2, maximum contact between cathode, oxygen and water is required due to which porous electrodes like gas- diffusion electrodes are used which provide high specific surface area. However, H_2O_2 is able to attack only some selective compounds like sulfur compounds, aldehydes, ketones, cyanides etc. due to which they are usually performed in the presence of Fe^{2+} ions to yield the Fenton's reagent. This process is called as the electro-Fenton process.

3.15.2 Electro-Fenton's Method

This is derived from the Fenton process and is an emerging technology for degradation of organics in wastewater [53]. The major steps involved in the mechanism of Electro-Fenton process are as follows (Eqs. 47-51)[54]:

- In situ electrogeneration of hydrogen peroxide through cathodic reduction of oxygen.
- Generation of hydroxyl radicals via Fenton's reaction between ferrous ions and electrogenerated H_2O_2
- Promoting the formation of hydroxyl radical physisorbed (anode(OH$^\bullet$)) on the electrode surface
- Regeneration of Fe^{3+}/Fe^{2+} by direct reduction on the cathode

$$O_2 + 2H^+ + 2e^- \rightarrow H_2O_2 \tag{47}$$

$$Fe^{2+} + H_2O_2 + H^+ \rightarrow Fe^{3+} + OH^- + OH^{\bullet} \tag{48}$$

$$Anode + H_2O \rightarrow Anode(OH^{\bullet}) + H^+ + e^- \tag{49}$$

$$Anode\ (OH^{\bullet}) + organic\ compounds \rightarrow Anode + oxidized\ products \tag{50}$$

$$Fe^{3+} + e^- \rightarrow Fe^{2+} \tag{51}$$

Apart from this, there are couple of other variants of Electro-fenton process which are sonoelectro-Fenton and bioelectro-Fenton processes [54]. In the sonoelectro-fenton process, the ultrasound is coupled with the electro-Fenton process for favouring the regeneration of Fe^{2+} from the complexes. Additionally, ultrasound plays a role in improving the efficiency through physical and chemical mechanisms. In the bioelectro-Fenton process, the cost effective biological methods are coupled to the electro-Fenton process.

3.16 Sulfate Radical based AOPs

$S_2O_8^{2-}$ (peroxydisulphate) radical is a strong oxidant with a standard oxidation potential of 2.01 eV. It can be activated by UV irradiation, heat, pH, or transition metals to form more powerful sufate radicals (Eqs. 52 and 53) which have an oxidation potential of 2.6 eV [5].

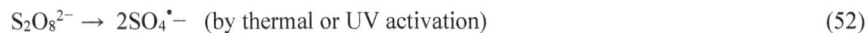

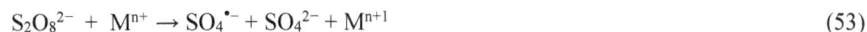

$$S_2O_8^{2-} \rightarrow 2SO_4^{\bullet-} \quad (by\ thermal\ or\ UV\ activation) \tag{52}$$

$$S_2O_8^{2-} + M^{n+} \rightarrow SO_4^{\bullet-} + SO_4^{2-} + M^{n+1} \tag{53}$$

Apart from this, sulphate radicals can also produce hydroxyl radicals as per the Eqs. 54 and 55 [55]:

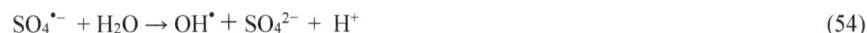

$$SO_4^{\bullet-} + H_2O \rightarrow OH^{\bullet} + SO_4^{2-} + H^+ \tag{54}$$

$$SO_4^{\bullet-} + OH^- \rightarrow OH^{\bullet} + SO_4^{2-} \qquad\qquad (55)$$

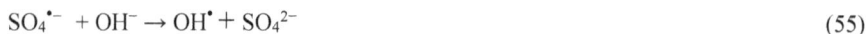

Many studies have reported AOPs based on sulphate radicals. Moussavi et al. [56] have compared VUV and UV/sulphate process for degradation and mineralization of cyanide with VUV process showing better results. Olmez-Hanci et al. [57] compared the AOPs using persulphate (PS), hydrogen peroxide and peroxymonosulphate (PMS) as free radical source for phenol degradation. Highest first order rate constant for degradation was shown by process with PMS thereby illustrating the fact that sulphate radical based AOPs can easily compete with hydroxyl radical based AOPs and the superior process of the two is dependent on the target pollutant and other operating variables. Additionally, there are some inherent advantages of sulphate bassed AOPs. Dewil et al. [51] have reported that sulphate based AOPs are pH independent and hence don't require the use of additional reagents to adjust the pH of the reaction medium unlike the hydroxyl based AOPs whose efficiency decreases with increasing pH. Moreover, longer lifetime of $SO_4^{\bullet-}$ radicals in water and reduced self scavenging effect as compared to OH radicals leads to higher mineralization efficiencies. Another beneficial feature in the form of higher selectivity of $SO_4^{\bullet-}$ permits destruction of specific functional groups in toxic compounds. However, sulphate based AOPs have certain limitations which probably have prevented its commercial growth. Use of metals like Cobalt for $SO_4^{\bullet-}$ generation and other sulphur based reagents mean that recovery of metals and toxicity associated with residual sulphates are a major concern. Despite great deal of studies, experiments in continuous modes and subsequent scale up to test their robustness are scarce.

Table 2 summarizes some of the studies of majority of the AOPs and their performance in degradation of different pollutants.

Table 2: Summary of various AOPs and their performance in the degradation of different types of pollutants.

Sr. No	Target pollutant/water matrix	AOP	Reaction conditions	Degradation/Mineralization	Ref.
1	Endotoxin Escherichia coli	O_3/H_2O_2	$[O_3]_0$ = 2 mg/L, $[O_3]_0/[H_2O_2]_0$ = 2, pH=7.4.	Degradation of 80%	[58]
		UV/H_2O_2	5 mg/L of H_2O_2 under 480 mJ/cm^2 (light intensity = 1 mW/cm^2 for 8 min) of monochromatic UV irradiation at 254 nm	Degradation of 68%	
2	Synthetic greywater	UV/O_3	Pilot plant system with 110 L of greywater circulated	Second-order rate constant for OH$^{\bullet}$	[59]

	containing orgnanics		through the reactor having 600W UV and 25g/hr ozone supply.	mineralization of greywater TOC was $[7.6 \pm 0.77]*10^7$ $M^{-1}s^{-1}$	
3	6-mercaptopurine (6-MP)	Heterogenous photocatalysis(TiO_2 +H_2O_2)	Under solar irradiation with optimal pH of 3.5 and H_2O_2 concentration of 3mM.	> 98% degradation	[60]
4	Amoxicillin and ciprofloxacin (hospital effluent)	UV-persulphate	Hg lamps with peak λ=254 nm as UV source. Optimal conditions: initial conc: 1 mg/L, persulphate conc: 10 mg/L, reaction time =60 min.	99.9% and 99.26% removal of ciprofloxacin and amoxicillin respectively. Negligible mineralization.	[61]
5	Dairy wastewater	Sono/persulphate	Optimal conditions: persulphate concentration =500mg/L, pH=3, US intensity=130kHz, contact time=60 min.	74.5 % COD removal	[62]
6	Ortho-Toluidine (OT)	UV/O_3/H_2O_2	Mercury lamp source for UV. Optimal conditions: H_2O_2=20mM, pH=10, ozone=0.5 L/min.	Almost 100% degradation in 40 min. 82.5 % mineralization in 120 min.	[63]
7	Quinolone antibiotic, norfloxacin	Gamma Radiation	Co-60 gamma-ray irradiation source. Optimal conditions: 97 Gyh^{-1} dose rate, N_2O saturation, low pH. Initial conc = 3.4 m/L.	98.6 % degradation efficiency	[64]
8	Semi-coke wastewater	SCWO	Optimal conditions of 600°C, 25 MPa, 1.3 times oxidation coefficient and 10 min.	COD removal: 99.02% NH_3-N removal: 63.94 %	[65]
9	Humic substance (HS) in wastewater	CWAO using $NiCo_2O_4$ (NCO) spinel	5 g of NCO for 1 L of HS solution. Optimal pH of 12 and temperature of 90 ^{0}C.	93.4 % reduction of TOC in 24 hr.	[66]
10	methylparaben, ethylparaben, propylparaben and butylparaben	Fenton	hydrogen peroxide and Fe^{2+} ion equal to $2.6*10^{-4}$ and $2.62*10^{-5}$M	93.0–99.8 % removal efficiencies	[67]

11	ciprofloxac in hydrochlor ide (CIP)	EF	Carbon felt cathode and Pt anode. Optimal current of 400 mA, Fe^{2+} concentration at 0.1 mM.	>94% mineralization in 6 hrs	[68]
12	Oxalic acid waste from chemical decontami nation of Nuclear Power Plant.	PF	A 120 W amalgam lamp emitting 254 nm. Fe^{2+} concentration of 2 mmol/L. Optimal ratio of 1:1 for concentration of Oxalic acid:H_2O_2.	Almost complete removal of O.A in 90 min.	[69]
13	Petroleum hydrocarbo ns (PHC) in oil spill sludge	US/Fento n (sono-Fenton)	Ultrasonic power of 100W with H_2O_2/Fe^{2+} weight ratio of 10:1 and ultrasonic treatment time of 10 min at pH of 3 sludge/water ratio of 1:100.	PHC removal rate of 62.99 %	[70]
14	Synthetic softdrink industry wastewater	MW photolysi s (MWP)	microwave oven reactor operating at a frequency of 2.45 GHz, 400 W with EDL.	chemical oxygen demand (COD), total Kjeldahl nitrogen (TKN) and total phosphate (TP) removal much higher in MWCP (7.48, 82.5 and 0.63 mg/L/min respectively).	[71]
		MW photocata lytic (MWPC)	microwave oven reactor operating at a frequency of 2.45 GHz with EDL and 3 g/L of Granular Activated Carbon supported TiO_2.		
15	Carbamazep ine, Tetracycline Hydrochlori de, Rhodamine B, Phenol and Thiamethox am	MW/UV	UV lamp (365 nm) with MW power of 800 W. Degradation studied with activated H_2O_2, sodium percarbonate (SPC), persulfate (PS), and peroxymonosulfate (PMS).	Synergy between UV and MW for degradation of all compounds. The order of a dominant radical generation rate was SPC(OH$^\bullet$)> H_2O_2(OH$^\bullet$)> PS (SO$_4^{\bullet-}$ > PMS (SO$_4^{\bullet-}$)	[72]

16	benzene (B), toluene (T), o-xylene (X), and naphthalene (N)	(MW-CWPO) using activated carbon (AC) as catalyst	100-mL stoppered PTFE reactor with AC at 1 gL^{-1} and H_2O_2 inside MW furnace. Intial aqueous pollutant solution (B:1.28 mM, T:1.09 mM, X:0.94 mM, and N: 0.78 mM).	Complete pollutant elimination with 90% TOC removal after 15 min.	[73]
17	4-chlorophenol (4-CP)	MW–H_2O_2	MW reactor of 400 W. Initial H_2O_2 concentration of 11 g/L and 4-CP concentration of 1000 mg/L.	Degradation efficiency of 34 % in 20 min at 180 ^{0}C.	[74]
18	Organophosphorus pesticides (chlorpyrifos and diazinon)	US	US of 20 kHz and 900W. 500 mL aqueous solution of chlorpyrifos (1.4 mg L^-^{1}) and diazinon (2.4 mg L^-^{1}). Reaction time of 60 min.	85% and 57% degradation of chlorpyrifos and diazinon respectively	[75]
19	Acetaminophen (ACP)	US	US of 600 kHz and 60W. ACP concentration: 82.69 $\mu molL^{-1}$	100 % degradation on 250 min and 30% TOC removal in 500 min.	[76]
20	phenol, bisphenol A (BPA) and diuron	US/photo-Fenton	US of 400 kHz and 450W. Treatment time of 40 min under solar radiation with 200 mg L^{-1} of H_2O_2 and 1 mg L^{-1} of Fe^{2+}	*Phenol Degradation*: 100% in 40 min TOC: 85% in 120 min *Diuron Degradation* Degradation: 40% in 300 min *BPA Degradation*: 100% in 60 min TOC: 8% in 300 min	[77]

4. Comparison of AOPs

Various types of AOPs, their mechanism along with their pros and cons were discussed in the above sections. The above mentioned studies give us only a glimpse of the tremendous potential of AOPs in degrading/mineralizing variety of organic pollutants in different types of wastewaters. Although the underlying mechanism along with their pros and cons give some insights into the applicability of these AOPs, it is difficult to choose one AOP over the other based on these studies. The sheer number of variables in all these studies (pollutant concentration, pH, temperature, oxidant dosage, light/radiation intensity and wavelength, reactor design, water matrix etc.) make it extremely difficult to compare the AOPs with one another. Although some studies have taken up more than one AOP in their experiments to compare their performances, more AOPs in a single study under similar set of conditions give deeper insights into the suitability of a particular AOP for a particular type of effluent/pollutant. Moreover, such an approach clearly lays down the preferred AOP for a particular wastewater which is vital for its commercial viability. Table 3 summarizes some of the comparison studies done.

Table 3: Summary of studies showing comparison between AOPs.

Pollutant/Effluent/Water Matrix		Parameter Studied	AOPs compared	AOP with best performance	Ref.
Landfill leachate		COD removal (BOD_5/COD ratio)	Photo-Fenton(UV), Fenton and UV-H_2O_2	Photo-Fenton(UV)	[78]
		COD removal, biodegradability and color removal	Homogenous photo-Fenton and heterogeneous photo-Fenton	Homogenous photo-Fenton	[79]
p-hydroxybenzoic acid		Degradation	UV, O_3, UV/TiO_0, O_3/Fe^{2+}, O_3/H_2O_2, O_3/UV, UV/H_2O_2, H_2O_2/ Fe^{2+}, H_2O_2/Fe^{2+}/O_3, UV/H_2O_2/O_3, H_2O_2/Fe^{2+}/UV and O_3/UV/H_2O_2/Fe^{2+}	O_3/UV/H_2O_2/Fe^{2+}	[80]
Endotoxin (E.Coli)		Degradation	O_3,UV,O_3/H_2O_2 and UV/H_2O_2	O_3/H_2O_2	[58]
Phenol		Degradation	O_3,O_3/H_2O_2,UV,UV/O_3,UV/H_2O_2,O_3/UV/H_2O_2,Fenton (Fe^{2+}/H_2O_2) and photocatalysis	Fenton (Fe^{2+}/H_2O_2)	[81]
		TOC removal (mineralization)	O_3,O_3/H_2O_2,UV/H_2O_2, UV/O_3 and UV/H_2O_2/O_3	UV/H_2O_2/O_3	[18]
N-methyl-2-pyrolidone (NMP)		TOC removal (mineralization)	UV/O_3, O_3 and UV/H_2O_2	UV/O_3	[82]
Herbicides	glyphosate (N-phosphonomethyl glycine)	TOC removal (mineralization)	O_3, photolysis and heterogenous photocatalysis (TiO_2)	O_3 at pH 10	[83]
	Phenylurea (fenuron, monuron and diuron)	TOC removal (mineralization)	UV, O_3, UV/O_3, TiO_2 (heterogenous photocatalysis) and UV/TiO_2	TiO_2 and UV/TiO_2 both gave significant mineralization. Rest didn't.	[84]
Petroleum/	Ortho-Toluidine (OT)	Degradation and Mineralization	UV/O_3/H_2O_2, UV/O_3, O_3/H_2O_2, and O_3	UV/O_3/H_2O_2	[63]

Petro chemi cal Waste water	Bitumen Effluent	COD and BOD reduction	O_3, H_2O_2 and O_3/H_2O_2 (Peroxone)	O_3/H_2O_2 (Peroxone)	[85]
	Oil spill sludge (nC_7–C_{10}, nC_{11}–C_{20} and nC_{21}–C_{30} fractions)	Degradation	Fenton, US and sono-Fenton	Sono-Fenton	[70]
	Cyanide	Degradation	VUV and UVC/$S_2O_8^{2-}$	VUV	[56]
	Alkyd resin wastewater	TOC removal (mineralization)	Fenton process, O_3/H_2O_2, electro-oxidation (EO), wet air oxidation and cavitation using ultrasonic horn (US)	O_3/H_2O_2 (cost effective)	[86]
Dyes	Reactive Orange 4 (RO4) and ReactiveYello w14 (RY14)	Decolorization	Photo-Fenton ($Fe^{2+}/H_2O_2/UV$), UV/TiO_2 and UV/H_2O_2	Photo-Fenton ($Fe^{2+}/H_2O_2/UV$)	[87]
	C.I. Reactive Red 45	TOC and AOX removal (mineralization)	UV/H_2O_2, UV/O_3, and UV/H_2O_2/O_3	UV/H_2O_2/O_3	[17]
	Ponceau S (PS)	Decolorization	H_2O_2/UV, Fenton (Fe^{2+}/H_2O_2), Fe^{2+}/H_2O_2/UV photo Fenton and solar photo-Fenton	$Fe^{2+}/H_2O_2/UV$ photo Fenton	[88]
Phar ma	carbamaze- pine (CBZ), diclofenac (DCF), and trimethoprim (TPM) and	Degradation	Heterogeneous photocatalysis (sunlight/N doped-TiO_2) and homogenous sunlight/H_2O_2	sunlight/N-doped TiO_2	[89]
		Degradation	H_2O_2/sunlight, solar photo-Fenton (Fe^{2+}/H_2O_2/sunlight), Fe^{2+}/H_2O_2/EDDS complex/sunlight and sunlight/N doped-TiO_2	$Fe^{2+}/H_2O_2/E$ DDS complex/sunl ight	[90]
	60 real effluents	COD removal	Fenton oxidation (FO) and conductive-diamond electro-oxidation (CDEO)	CDEO	[91]

It can be seen that, in general, hybrid AOPs utilizing the beneficial aspects of two or more individual AOPs show best results. Therefore, standalone processes using just O_3, H_2O_2 etc. are generally not preferred. However, merely combining the AOPs do not necessarily make them cost effective. Hence, some studies consisting of cost analysis of various AOPs give a better idea about the suitability of a particular AOP for a particular type of effluent. For e.g.: for the landfill leachate, the Fenton treatment has been reported to be the most cost effective AOP [92]. Similar cost studies done by Krishnan et al. [93]

have suggested that H_2O_2/O_3 and H_2O_2/UV are more economical as compared to photocatalytic (TiO_2/UV) process and Fenton's reactions.

Barndõk et al. [94] have studied the degradation of 1,4-Dioxane-containing wastewater from the chemical industry. A detailed cost analysis was done based on the lab scale trials where the target was 40% reduction in COD for a wastewater treatment capacity of 43800 $m^3 \cdot y^{-1}$. The results suggested that ozonation (O_3), conductive diamond electrochemical oxidation (CDEO), and zero valent iron-based heterogeneous photo-Fenton processes (both solar and UV) all displayed similar magnitude of treatment cost of about ≈ 5 €/m^3. This again highlights the fact that pilot plant studies are essential to assess the actual commercial viability of any process.

Since cost is also related to the energy consumption of an AOP, some studies have used the term Electrical Energy per Order (EEO) for evaluating the energy efficiencies of different AOPs. EEO is given by the formula:

$$EEO = \frac{P_{el}*t*1000}{V*60*log\left(\frac{Ci}{Cf}\right)}$$

Where, electrical energy per order (kWh/m^3order[1]), P_{el} is the electrical power input (kW), t is the irradiation time (min), V is the volume of effluent used (L), C_i and C_f are the initial and final effluent concentration (ppm) [95]. EEO depends on factors like rate constants, concentration range of pollutant (only if > 1mg/L), molecular structure etc. Moreover, factors like water matrix, capacity, chemical dosage etc. can also have influence on the overall efficiency of the process. So, optimization of a particular AOP with respect to reactor geometry, oxidant dosage etc. is a must before evaluation of EEO [7].

Rosenfeldt et al. [96] have considered oxidant and chemicals used (H_2O_2) as stored electrical energy. They studied the degradation of pCBA for comparing four AOPs (ozone, ozone+H_2O_2, low pressure UV (LP)+H_2O_2, and medium pressure UV (MP)+H_2O_2). Their results revealed that the ozone-based processes were the most energy efficient ones than the UV/H_2O_2 based processes. Further, process with equimolar addition of H_2O_2 in ozone process was found to be the most energy efficient. Interestingly, high operational costs of UV based AOPs were mostly due to the cost of H_2O_2 rather than the operating cost of UV lamps. On the other hand, Asaithambi et al. [95] have reported that ozone/UV based Fenton process was the most energy efficient (among O_3/UV, $O_3/UV/H_2O_2$, $O_3/UV/Fe^{2+}$, O_3, O_3/Fe^{2+}, UV/H_2O_2, $UV/H_2O_2/Fe^{2+}$ and $O_3/UV/Fe^{2+}/H_2O_2$) along with the best removal performance of color and COD in distillery wastewater. Muruganandham et al. [87] have carried out similar study where

EEO based comparison revealed that UV Fenton process was more effective in degradation of azo dyes (Reactive Yellow and Reactive Orange) as compared to UV/H_2O_2 and UV/TiO_2.

Miklos et al. [7] have carried out a comprehensive comparison of various AOPs based on their EEO values (Fig. 5). They further classified the AOPs into three groups based on their median values of EEOs.

Processes with median EEO values <1 kWh/m^3 ($O_3,O_3/H_2O_2,O_3/UV,UV/H_2O_2$, UV/persulfate, UV/chlorine and electron beam) are categorized under group 1 which indicate their great potential for large scale application. Photo-Fenton, plasma and electrochemical AOPs with median EEO values of 2.6, 3.3, and 38.1 kWh/m^3 (1-100 kWh/m^3), respectively, are energy intensive for practical purposes but might find usefulness when applied for specific targets. Processes like UV-based photocatalysis, ultrasound and microwave-based AOPs are highly energy intensive and are not viable commercially with median EEO values > 100 kWh/m^3.

Some obvious points can be taken into consideration for deciding on the suitability of an AOP for treating a particular type of effluent. For example, the major limitation of Fenton process is the maintenance of pH at 3 during treatment. This might suggest that Fenton process is better suited and more cost effective for acidic wastewaters. Similarly, effluents with intense color might be kept away from AOPs using light as light penetration to deeper layers would not be satisfactory leading to a less effective process [97]. The above studies surely indicate that, despite the attractive results, significant number of AOPs fall on wrong side of the current cost effectiveness status. However, extensive research is being carried out in some of these fields which may render them suitable for practical applications sooner than later. Table 4 summarizes some of the main advantages and limitations of various AOPs discussed in literature [98].

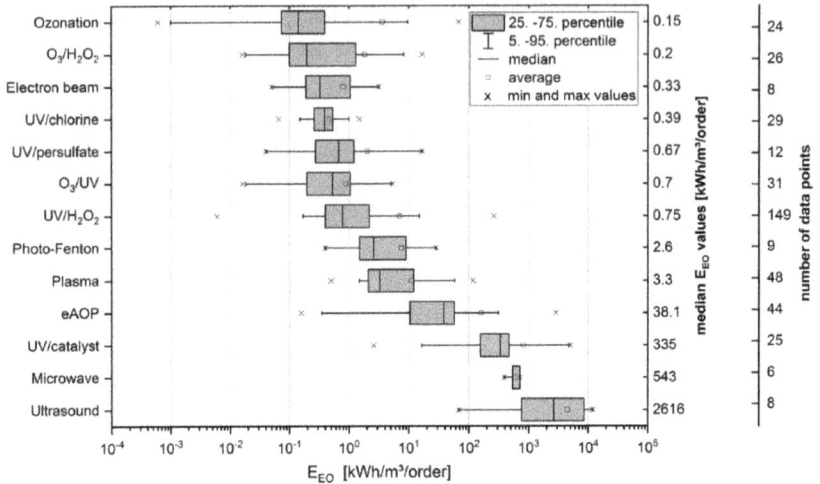

Figure 5: Overview of published EEO-values of different AOPs sorted according to median values (Reprinted with permission from Ref. [7]. Copyright 2018, Elsevier).

Table 4: Advantages and Limitations of AOPs.

AOP	Advantages	Disadvantages
O_3	• Effective for a wide range of contaminants • Existence of many full scale plants	• Risks associated with ozone generation • Reduced ammonia removal
O_3/H_2O_2	• More hydroxyl radical generation as compared to O_3	• Excess residual peroxide can create problems in assessment of removal of pollutant
UV/O_3 UV/H_2O_2 $UV/O_3/H_2O_2$	• Degradation of organics can occur by both radical reactions as well as photolysis by UV • UV light can result in disinfection by killing microbes. • Quantities of ozone and peroxide can be optimized depending on the target	• Excess peroxide can lead to hydroxyl radical scavenging • High operating costs associated with UV lamps. • Increased turbidity may prevent UV light to reach greater depths which will

		reduce the overall efficiency.
Fenton based AOPs	• Less energy requirements • Usage of non hazardous reagents and easy reactor design	• Requirement of acidic conditions • Problems associated with sludge generation
Electrochemical Oxidation	• No additional chemicals required • Final products are non hazardous in nature • High efficiency	• Excessive energy requirement
Ultrasound based	• Versatile technology which can be integrated with other AOPs as well. • Suitable for small volumes.	• High energy consumption (high EEO values) • Sonotrode erosion issues
Microwave based	• Increased selectivity for target specific applications • Reduction in reaction time and activation energy.	• High energy consumption (high EEO values)
Electron Beam	• No requirement of chemical reagents • Can do both disinfection and degradation of pollutants.	• High Operational costs
SCWO	• Can be used for highly concentrated wastewaters (1-20% organic matter) • Possibility of energy recovery in the system	• High initial investment required • High maintenance and repair costs for equipments working under extreme conditions
WAO/CWAO	• Mild conditions are involved without need for any hazardous reagent • Can be integrated with other biological processes for higher efficiencies and cost effectiveness.	• Deactivation of heterogenous catalysts due to metal leaching and carbonaceous deposits • Complete oxidation to CO_2 and H_2O is difficult to achieve
Sulphate based AOPs	• pH independent process • longer lifetimes of sulphate radicals as compared to hydroxyl radicals	• Residual sulphates may pose a problem • Lack of scale-up studies for prospects in commercial viability

Heterogeneous Photocatalysis	• Ability to work under ambient conditions • Potential to use sunlight through visible light active photocatalysts • Use of non toxic and inexpensive inputs	• Existing reliance on UV lamps makes it less economical • Problems associated with recovery and reuse of photocatalyst from treated water.

5. Commercialization/practical application of AOPs

As discussed in the earlier sections, many laboratory studies are available in literature which show great pollutant removal abilities and promise to be a better alternative to the existing treatment plants. Comparison of various AOPs suggest that for a particular target pollutant, one AOP maybe better suited and is more cost effective than the others. Also, AOPs like UV based photocatalysis, ultrasound and microwave based processes are quite energy demanding and hence their commercial viability is still to be worked upon despite their enhanced performances in lab studies.

Ozone based processes were some of the first AOPs to be applied practically in full scale. An ozone/UV system was set up in Tinker Air Force Base (Oklahoma, USA) to treat cyanides and refractor organics. In 1978, an ozone treatment facility was installed in Cadillac Motor Car Division of General Motors Corporation in Detroit, Michigan (USA). Other commercial applications include the one at Bell Telephone Laboratories in 1973 for the control of bacteria, ozone/GAC system at ARCO Products Company, Richmond, California, USA where treatment of oily wastewater is being done since 1991 [9].

Similarly, a $O_3/H_2O_2/UV$ system was erected for the removal of VOCs in groundwater at San Jose, USA [9]. Khan et al. [99] have reported how AOPs have been used for the supply of potable water. AOP plants, especially those with UV/H_2O_2 process have been operating in many places around the world. Some of them include the Groundwater Replenishment System in Orange county (California, USA) with a 350 ML/day capacity implemented in 2008, Raw Water Production facility in Big Spring (Texas, USA), UV/H_2O_2 AOP plant in Beufort West Municipality, South Africa since 2011 with 2 ML/day capacity. For removal of micro-pollutants, a plant based on UV/H_2O_2 treatment has been installed at Andijk, The Netherlands. 80% atrazine degradation is achieved with a UV dose of 5.4 kJ m−2, 0.006 gL−1 H_2O_2 and has a capacity of 95000 m^3/day [100].

As far as application of Wet Air Oxidation is concerned, "The Zimpro TSC process" was widely used in the earlier days in the USA and Europe. Later on, homogenous processes like The VerTech process, The LOPROX process developed by Bayer AG, The WPO process developed in France etc. were used in different parts of the world. Various heterogeneous WAO processes like The Osaka Gas process (based on precious and base metals like titania and zirconia for treatment of industrial and urban wastes) and The Kurita Process (for ammonia removal) were developed. These systems with differing conditions were used for treating wastewaters from different sources [101].

Ultrasound based AOPs are also catching up in the commercial aspect. These can be basically operated as batch or flow types. In the flow types, FFR Ultrasonics Ltd. developed a high power system where sonicated disc inserts within a tube were used. 10 such discs, 3kW of power each, were connected in series making it capable of handling 33L of fluid every second [102].

Solar based plants have been installed at different locations for the treatment of industrial wastewater. AOPs involving light in the form of UV have always longed for a system which is capable of working under sunlight. The huge reductions in the operating costs due to UV lamps have made solar based AOPs one of the most attractive methods of pushing AOPs like photo-Fenton and heterogenous photocatalysis towards commercial feasibility. Solar photocatalysis has been used for the treatment of landfill leachate of landfill site in Rethymnon, Crete, Greece [100].

A solar-Fenton unit combined with biological treatment was installed in Spain (Plataforma Solar de Almeria) featuring a CPC (compound parabolic collector) technology. This was part of the SOLARDETOX project in Europe for treating recalcitrant pollutants [103]. Later, a new CPC based plant was erected by ALBIADA company (Fig. 6) which mineralizes 80% of TOC in batch process. A pilot plant based on combined electron beam and biological purification process was developed for treating dye wastewater at Taegu City, South Korea in 1998. This system can treat 1000m³/day of wastewater. The facility uses a 1MeV, 40kW ELV electron accelerator manufactured by Institute of Nuclear Physics, Siberian Division of Russian Academy of Sciences, Novosibirsk. Wang et al. [104] have reported that a recent commercial plant for treatment of industrial wastewater using electron beam radiation was established in 2017 in Jinhua, Zhejiang Province, China. This facility for treating dye wastewater operates with a capacity of 2000 m³/day.

In a nutshell, AOPs are gradually catching the attention of the scientific community and governing bodies and a significant number of AOP based treatment plants have been installed across the globe. Although, some AOPs are not yet cost effective for commercial viability, extensive research on AOPs and novel ways of reducing the capital and operating costs are being sought in recent years. This promises to take them one step closer at a time towards practical application.

Figure 6: A Solar detoxification demonstration plant erected by ALBIADA in Spain (Reprinted with permission from Ref. [103]. Copyright 2007, Elsevier).

6.   Recent developments

Extensive research over the last few years have resulted in some novel ideas which further justify the beneficial aspects of AOPs for wastewater applications. One of the core ideas based on which recent research is moving forward is the idea of synergism where one or more AOPs are integrated together in a novel way. In the previous sections, some of such combinations were already discussed (like $UV/H_2O_2/O_3$, electro-Fenton, US/UV etc.) where the combination of AOPs generally have led to enhanced efficiencies for pollutant removal in the lab studies. Recent studies have focused more on this aspect where novel hybrids between AOPs or between AOP and other treatment methods have been explored.

In general, synergy is defined by the following formula:

$$S = \frac{k_{combined} - \sum_1^n k_i}{k_{combined}}$$

Where, $k_{combined}$ is the rate constant obtained in the combined AOP process and $\sum_1^n k_i$ represents the summation of rate constants in the standalone individual AOPs. A positive value for S represents synergistic effect and vice versa [51]. Table 5 summarizes some recent novel hybrid AOPs.

Table 5: Hybrid AOPs and their degradation performance.

Hybrid AOP	Pollutant/Water Matrix	Performance	Ref.
solar photoelectro- Fenton (SPEF) coupled with solar heterogeneous photocatalysis (SPC)	Salicylic Acid	87% mineralization, 13% current efficiency and 1.133 kWhg^{-1} TOC energy consumption (6.0 AhL^{-1}) after 360 min of electrolysis	[105]
Electrocoagualtion coupled with Electro-Fenton	Textile industry wastewater	Color, turbidity and TOC removal of 100%, 100% and 97% respectively.	[106]
sequencing batch biofilter granular reactor (SBBGR) coupled with solar photo-Fenton (SphF)	Municipal landfill leachate	For target COD of 160 mg/L, hybrid setup was more economical than the standalone setups.	[107]
supercritical Fenton oxidation (SCFO)	P-aminonaphthalenesulfonic Acid	Superior degradation as compared with the SCWO	[108]
sequencing batch reactor (SBR) coupled with electro-Fenton	Pharmaceutical wastewater	Enhanced COD removal and Complete degradation of pharmaceuticals.	[109]
Photo Electro Fenton Process combined with membrane bioreactor	Landfill leachate	Enhanced removal of TSS, BOD, COD, Ammonia Nitrogen, Phosphate, Sulphate, Sulphide and Chloride.	[110]
electro-assisted CWAO (ECWAO) using partially oxidized nickel (Ni@NiO) immobilized on a porous graphite felt (GF) as a catalytic anode	triclosan (TCS), bisphenol A, ciprofloxacin, sulfamethoxazole, congo red, crystal violent and rodamine B	High efficiency in mineralization at low energy consumption	[111]
Ultrasound assisted heterogeneous photocatalysis	Vinasse	Highest removal efficiencies of Polyphenols, COD, and color in hybrid as compared to standalone processes	[112]
TiO$_2$-based photocatalysis combined with photo-Fenton process	Formaldehyde	Synergy between the two AOPs without the need of H$_2$O$_2$ addition.	[113]

Materials Research Forum LLC
https://doi.org/10.21741/9781644901144-2

Paździor et al. [97] have reviewed many AOPs combined with biological treatment methods for the treatment of industrial textile wastewater. Recently, heterogenous photocatalysis has picked a great deal of momentum with concepts like Photocatalytic Membrane Reactors (PMRs) where membrane separation is integrated with photocatalysis. Studies dealing with photocatalyst in slurry form or in an immobilized form have been carried out with PMRs which offer another promising angle to tackle photocatalyst recovery problem which is one of the major limitations of photocatalysis [114]. Furthermore, novel photocatalysts have been developed in the recent years like magnetic photocatalysts [115] and floating photocatalysts [116] which again resolve the photocatalyst recovery problems. These ideas when coupled with highly effective photocatalysts like black TiO_2 [117], which are visible light active, can completely eliminate all drawbacks of heterogenous photocatalysis and hence can make the system cost effective and operable under sunlight. Moreover, heterogenous photocatalysis is probably the most researched AOP currently. So, its highly likely that more large scale plants based on photocatalytic wastewater treatment are seen across the world in the coming years.

7. Summary and future challenges

It is quite clear that Advanced Oxidation Processes have revolutionized the research for wastewater treatment. With the current global water crisis, AOPs have offered a promising alternative to the conventional water treatment techniques. Many different AOPs have been reported till date for removal of various types of contaminants from different types of effluents/water matrices. Starting from the simple ozonation to the sophisticated AOPs based on electron beam, ultrasound and microwave, tremendous research and development can be seen in the last three to four decades. It is obvious that all these AOPs have their own set of advantages and limitations and have their own areas of cost effective applicability. Hybrid AOPs are explored for creating synergistic effects between two or more different systems for higher efficiencies as well as cost effectiveness. Recent studies done in this line promise to make AOPs much more commercially viable in the years to come. Nevertheless, AOPs have made decent progress till date boasting good figures when it comes to number of installed AOP based commercial plants across the globe. This number can be further improved by keeping pace with the advancements in the lab scale research.

However, some challenges do exist for the practical applications of AOPs. For example, for the disinfection process, chlorination is still the most cost-effective method compared to various AOPs [118]. So, further research on this area is needed to eliminate the problem of reduced lifetime of free radicals which is one of the main reasons for reduced performance in disinfection [1]. Similarly, AOPs have been also proven to be less cost effective for landfill leachate. Ammonia removal by $OH^\bullet$ radicals is quite slow and also the performance of AOP is affected by the presence of organics in the actual wastewater [119]. Thus, research studies based on actual wastewaters are vital as lab scale degradation of a single pollutant using AOPs does not give reliable results for scale-up.

What is also clear is that AOPs in most cases need pretreatment in the form of primary, secondary, tertiary treatments or combination of one or more such strategies. There might be specific applications where adsorption using activated carbon will be more cost effective than AOPs. The key in such cases is to look at the possibility of integrating AOPs with existing facilities so that the combination becomes even more cost effective than the existing conventional method. Hence, instead of completely replacing the existing technology, AOPs can rather be used as an additional facility to further polish the already treated water.

Understanding the degradation mechanisms is crucial for AOP studies and its subsequent practical implementation. Studies involving the intermediates formed after initial degradation and the underlying mechanism are the most difficult but vital from the toxicity point of view. Hence, getting deeper insights into the degradation mechanisms is essential before application to real life problems.

Currently, utilization of solar radiation in wastewater applications is gaining momentum. AOPs like heterogenous photocatalysis are probably the future of water treatment mainly because of massive global research involved in developing visible light active photocatalysts which are easily recoverable and can be regenerated on site. Although, such photocatalysts will take some time to enter the market, they certainly do offer interesting prospective. Meanwhile, the extra UV radiation reaching the earth's surface through sunlight due to ozone layer depletion can be put to good use in relevant areas in AOPs reliant on UV light.

Overall, AOPs have been around for some decades now and have shown promising results in terms of becoming an alternative treatment strategy during this global water crisis. However, certain aspects in AOPs still need more attention from research as well as scale-up point of view. Playing to the strengths is going to be the key for AOP based water treatment operations currently. The future of AOPs does look bright as many limitations are being overcome using hybrid AOP technologies. With proper awareness and utilization of renewable energy sources, AOP can indeed become the future of wastewater treatment.

References

[1]   S.C. Ameta, Introduction, in: S. C. Ameta, R. Ameta (Eds.), Advanced Oxidation Processes for Wastewater Treatment: Emerging Green Chemical Technology, Academic Press, 2018, pp. 1-12. https://doi.org/10.1016/B978-0-12-810499-6.00001-2.

[2]   N.S. Topare, S.J. Attar, M.M. Manfe, Sewage / Wastewater treatment technologies : a review, Scietific Rev. Chem. Commun. 1 (2011) 18–24.

[3]   M. Muruganandham, R.P.S. Suri, S. Jafari, M. Sillanpää, G.J. Lee, J.J. Wu, M. Swaminathan, Recent developments in homogeneous advanced oxidation processes for water and wastewater treatment, Int. J. Photoenergy. 2014 (2014) 821674. https://doi.org/10.1155/2014/821674.

[4]   N. Azbar, Comparison of various advanced oxidation processes and chemical treatment methods for COD and color removal from a polyester and acetate fiber dyeing effluent, Chemosphere 55 (2004) 35–43. https://doi.org/10.1016/j.chemosphere.2003.10.046.

[5]   Y. Deng, R. Zhao, Advanced Oxidation Processes (AOPs) in Wastewater Treatment, Curr. Pollut. Reports. 1(3) (2015) 167–176. https://doi.org/10.1007/s40726-015-0015-z.

[6]   J.H. Carey, An introduction to advanced oxidation processes (AOP) for destruction of organics in wastewater, Water Poll. Res J. Canada. 27 (2018) 43. https://doi.org/10.1017/CBO9781107415324.004.

[7]   D.B. Miklos, C. Remy, M. Jekel, K.G. Linden, J.E. Drewes, U. Hübner, Evaluation of advanced oxidation processes for water and wastewater treatment – A critical review, Water Res. 139 (2018) 118–131. https://doi.org/10.1016/j.watres.2018.03.042.

[8]   S. Sharma, J. Ruparelia, M. Patel, A general review on advanced oxidation processes for waste water treatment, Int. Conf. Curr. Trends Technol. (2011) 8–10. http://nuicone.org/site/common/proceedings/Chemical/poster/CH_33.pdf.

[9]   R. Munter, Advanced Oxidation Processes- current status and prospects, Proc. Est. Acad. Sci. Chem. 50 (2011) 59–80. https://doi.org/10.1016/B978-0-444-53199-5.00093-2.

[10]   K. Ikehata, Y. Li, Ozone-Based Processes, in: S. C. Ameta, R. Ameta (Eds.), Advanced Oxidation Processes for Wastewater Treatment: Emerging Green Chemical

Technology, Academic Press, 2018, pp. 115–134. https://doi.org/10.1016/B978-0-12-810499-6.00005-X.

[11] A. Matilainen, M. Sillanpää, Removal of natural organic matter from drinking water by advanced oxidation processes, Chemosphere. 80 (2010) 351–365. https://doi.org/10.1016/j.chemosphere.2010.04.067.

[12] N. Mishra, R. Reddy, A. Kuila, A. Rani, A. Nawaz, S. Pichiah, A Review on Advanced Oxidation Processes for Effective Water Treatment, Curr. World Environ. 12 (2017) 469–489. https://doi.org/10.12944/cwe.12.3.02.

[13] J. Groele, J. Foster, Hydrogen Peroxide Interference in Chemical Oxygen Demand Assessments of Plasma Treated Waters, Plasma. 2 (2019) 294–302. https://doi.org/10.3390/plasma2030021.

[14] M. I. Stefan, UV photolysis: background , in: S. Parsons (Ed.), Advanced oxidation processes for water and wastewater treatment, IWA publishers, 2018, pp. 7–48. https://doi.org/10.1007/s11356-018-3411-2.

[15] J.C. Mierzwa, R. Rodrigues, A.C.S.C. Teixeira, UV-Hydrogen Peroxide Processes, in: S. C. Ameta, R. Ameta (Eds.), Advanced Oxidation Processes for Wastewater Treatment: Emerging Green Chemical Technology, Academic Press, 2018. pp. 13-48. https://doi.org/10.1016/B978-0-12-810499-6.00002-4.

[16] T. A. Tuhkanen, UV/H2O2 processes, in: S. Parsons (Ed.), Advanced oxidation processes for water and wastewater treatment, IWA publishers, 2018, pp.86-110, https://doi.org/10.1007/s11356-018-3411-2.

[17] I. Peternel, N. Koprivanac, H. Kusic, UV-based processes for reactive azo dye mineralization, Water Res. 40 (2006) 525–532. https://doi.org/10.1016/j.watres.2005.11.029.

[18] H. Kusic, N. Koprivanac, A.L. Bozic, Minimization of organic pollutant content in aqueous solution by means of AOPs: UV- and ozone-based technologies, Chem. Eng. J. 123 (2006) 127–137. https://doi.org/10.1016/j.cej.2006.07.011.

[19] C. Walling, Fenton's Reagent Revisited, Acc. Chem. Res. 8 (1975) 125–131. https://doi.org/10.1021/ar50088a003.

[20] R. Ameta, A.K. Chohadia, A. Jain, P.B. Punjabi, Fenton and Photo-Fenton Processes, in: S. C. Ameta, R. Ameta (Eds.), Advanced Oxidation Processes for Wastewater Treatment: Emerging Green Chemical Technology, Academic Press, 2018, pp. 49-87. https://doi.org/10.1016/B978-0-12-810499-6.00003-6.

[21]   M. hui Zhang, H. Dong, L. Zhao, D. xi Wang, D. Meng, A review on Fenton process for organic wastewater treatment based on optimization perspective, Sci. Total Environ. 670 (2019) 110–121. https://doi.org/10.1016/j.scitotenv.2019.03.180.

[22]   J.H. Sun, S.P. Sun, M.H. Fan, H.Q. Guo, L.P. Qiao, R.X. Sun, A kinetic study on the degradation of p-nitroaniline by Fenton oxidation process, J. Hazard. Mater. 148 (2007) 172–177. https://doi.org/10.1016/j.jhazmat.2007.02.022.

[23]   H. Hansson, F. Kaczala, M. Marques, W. Hogland, Photo-Fenton and Fenton oxidation of recalcitrant industrial wastewater using nanoscale zero-valent iron, Int. J. Photoenergy. 2012 (2012) 531076. https://doi.org/10.1155/2012/531076.

[24]   M. Tokumura, M. Sekine, M. Yoshinari, H.T. Znad, Y. Kawase, Photo-Fenton process for excess sludge disintegration, Process Biochem. 42 (2007) 627–633. https://doi.org/10.1016/j.procbio.2006.11.010.

[25]   D. Beydoun, R. Amal, G. Low, S. McEvoy, Role of Nanoparticles in Photocatalysis, J. Nanoparticle Res. 1 (1999) 439–458. https://doi.org/10.1023/a:1010044830871.

[26]   X. Yan, Y. Li, T. Xia, Black Titanium Dioxide Nanomaterials in Photocatalysis, Int. J. Photoenergy. 2017 (2017) 8529851. https://doi.org/10.1155/2017/8529851.

[27]   L. V. Bora, R.K. Mewada, Visible/solar light active photocatalysts for organic effluent treatment: Fundamentals, mechanisms and parametric review, Renew. Sustain. Energy Rev. 76 (2017) 1393–1421. https://doi.org/10.1016/j.rser.2017.01.130.

[28]   A. Talebian, M.H. Entezari, N. Ghows, Complete mineralization of surfactant from aqueous solution by a novel sono-synthesized nanocomposite (TiO_2-Cu_2O) under sunlight irradiation, Chem. Eng. J. 229 (2013) 304–312. https://doi.org/10.1016/j.cej.2013.05.117.

[29]   M. Sboui, M.F. Nsib, A. Rayes, M. Swaminathan, A. Houas, TiO_2–PANI/Cork composite: A new floating photocatalyst for the treatment of organic pollutants under sunlight irradiation, J. Environ. Sci. 60 (2017) 3–13. https://doi.org/10.1016/j.jes.2016.11.024.

[30]   W.J. Chung, D.D. Nguyen, X.T. Bui, S.W. An, J.R. Banu, S.M. Lee, S.S. Kim, D.H. Moon, B.H. Jeon, S.W. Chang, A magnetically separable and recyclable Ag-supported magnetic TiO_2 composite catalyst: Fabrication, characterization, and photocatalytic activity, J. Environ. Manage. 213 (2018) 541–548. https://doi.org/10.1016/j.jenvman.2018.02.064.

[31]   Q. Gao, Z. Liu, FeWO$_4$ nanorods with excellent UV–Visible light photocatalysis, Prog. Nat. Sci. Mater. Int. 27 (2017) 556–560. https://doi.org/10.1016/j.pnsc.2017.08.016.

[32]   R. Xiao, Z. Wei, D. Chen, L.K. Weavers, Kinetics and mechanism of sonochemical degradation of pharmaceuticals in municipal wastewater, Environ. Sci. Technol. 48 (2014) 9675–9683. https://doi.org/10.1021/es5016197.

[33]   R.A. Torres-Palma, E.A. Serna-Galvis, Sonolysis, in: S. C. Ameta, R. Ameta (Eds.), Advanced Oxidation Processes for Wastewater Treatment: Emerging Green Chemical Technology, Academic Press, 2018, pp.177-213. https://doi.org/10.1016/B978-0-12-810499-6.00007-3.

[34]   C.G. Joseph, G. Li Puma, A. Bono, D. Krishnaiah, Sonophotocatalysis in advanced oxidation process: A short review, Ultrason. Sonochem. 16 (2009) 583–589. https://doi.org/10.1016/j.ultsonch.2009.02.002.

[35]   M.P. Rayaroth, U.K. Aravind, C.T. Aravindakumar, Sonochemical degradation of Coomassie Brilliant Blue: Effect of frequency, power density, pH and various additives, Chemosphere. 119 (2015) 848–855. https://doi.org/10.1016/j.chemosphere.2014.08.037.

[36]   P. Verma, S.K. Samanta, Microwave-enhanced advanced oxidation processes for the degradation of dyes in water, Environ Chem Lett. 16 (2018) 969-1007. https://doi.org/10.1007/s10311-018-0739-2.

[37]   A. Chavoshani, M.M. Amin, G. Asgari, A. Seidmohammadi, M. Hashemi, Microwave/Hydrogen Peroxide Processes, in: S. C. Ameta, R. Ameta (Eds.), Advanced Oxidation Processes for Wastewater Treatment: Emerging Green Chemical Technology, Academic Press, 2018, pp. 215-255. https://doi.org/10.1016/B978-0-12-810499-6.00008-5.

[38]   N. Remya, J.G. Lin, Current status of microwave application in wastewater treatment-A review, Chem. Eng. J. 166 (2011) 797–813. https://doi.org/10.1016/j.cej.2010.11.100.

[39]   C.J. Jou, Degradation of pentachlorophenol with zero-valence iron coupled with microwave energy, J. Hazard. Mater. 152 (2008) 699–702. https://doi.org/10.1016/j.jhazmat.2007.07.036.

[40]   F. Parolin, U.M. Nascimento, E.B. Azevedo, Microwave-enhanced UV/H$_2$O$_2$ degradation of an azo dye (tartrazine): Optimization, colour removal, mineralization and ecotoxicity, Environ. Technol. (United Kingdom). 34 (2013) 1247–1253.

https://doi.org/10.1080/09593330.2012.744431.

[41]   M.D. Bermejo and M. J.Cocero, Supercritical Water Oxidation: A Technical
       Review, AIChE J. 52 (2006) 3933–3951. https://doi.org/10.1002/aic.

[42]   V. Vadillo, J. Sánchez-Oneto, J.R. Portela, E.J. Martínez de la Ossa, Supercritical
       Water Oxidation, in: S. C. Ameta, R. Ameta (Eds.), Advanced Oxidation Processes for
       Wastewater Treatment: Emerging Green Chemical Technology, Academic Press,
       2018, pp. 333–358. https://doi.org/10.1016/B978-0-12-810499-6.00010-3.

[43]   Y. Gong, Y. Guo, S. Wang, W. Song, Supercritical water oxidation of
       Quinazoline: Effects of conversion parameters and reaction mechanism, Water Res.
       100 (2016) 116–125. https://doi.org/10.1016/j.watres.2016.05.001.

[44]   V. Vadillo, M.B. García-Jarana, J. Sánchez-Oneto, J.R. Portela, E.J.M. de la Ossa,
       Supercritical water oxidation of flammable industrial wastewaters: Economic
       perspectives of an industrial plant, J. Chem. Technol. Biotechnol. 86 (2011) 1049–
       1057. https://doi.org/10.1002/jctb.2626.

[45]   M. Trojanowicz, K. Bobrowski, T. Szreder, A. Bojanowska-Czajka, Gamma-ray,
       X-ray and Electron Beam Based Processes, in: S. C. Ameta, R. Ameta (Eds.),
       Advanced Oxidation Processes for Wastewater Treatment: Emerging Green Chemical
       Technology, Academic Press, 2018, pp. 257-331. https://doi.org/10.1016/B978-0-12-
       810499-6.00009-7.

[46]   A. Vahdat, S.H. Bahrami, M. Arami, A. Bahjat, F. Tabakh, M. Khairkhah,
       Decoloration and mineralization of reactive dyes using electron beam irradiation, Part
       I: Effect of the dye structure, concentration and absorbed dose (single, binary and
       ternary systems), Radiat. Phys. Chem. 81 (2012) 851–856.
       https://doi.org/10.1016/j.radphyschem.2012.03.005.

[47]   P. Gehringer, H. Eschweiler, The use of radiation-induced advanced oxidation for
       water reclamation, Water Sci. Technol. 34 (1996) 343–349.
       https://doi.org/10.1016/S0273-1223(96)00763-9.

[48]   V.S. Mishra, V. V. Mahajani, J.B. Joshi, Wet Air Oxidation, Ind. Eng. Chem. Res.
       34 (1995) 2–48. https://doi.org/10.1021/ie00040a001.

[49]   K.H. Kim, S.K. Ihm, Heterogeneous catalytic wet air oxidation of refractory
       organic pollutants in industrial wastewaters: A review, J. Hazard. Mater. 186 (2011)
       16–34. https://doi.org/10.1016/j.jhazmat.2010.11.011.

[50]   J. Fu, G.Z. Kyzas, Wet air oxidation for the decolorization of dye wastewater: An

overview of the last two decades, Chinese J. Catal. 35 (2014) 1–7.
https://doi.org/10.1016/S1872-2067(12)60724-4.

[51]   R. Dewil, D. Mantzavinos, I. Poulios, M.A. Rodrigo, New perspectives for Advanced Oxidation Processes, J. Environ. Manage. 195 (2017) 93–99. https://doi.org/10.1016/j.jenvman.2017.04.010.

[52]   F.C. Moreira, R.A.R. Boaventura, E. Brillas, V.J.P. Vilar, Electrochemical advanced oxidation processes: A review on their application to synthetic and real wastewaters, Appl. Catal. B Environ. 202 (2017) 217–261. https://doi.org/10.1016/j.apcatb.2016.08.037.

[53]   P. V. Nidheesh, R. Gandhimathi, Trends in electro-Fenton process for water and wastewater treatment: An overview, Desalination. 299 (2012) 1–15. https://doi.org/10.1016/j.desal.2012.05.011.

[54]   V. Poza-Nogueiras, E. Rosales, M. Pazos, M.Á. Sanromán, Current advances and trends in electro-Fenton process using heterogeneous catalysts – A review, Chemosphere. 201 (2018) 399–416. https://doi.org/10.1016/j.chemosphere.2018.03.002.

[55]   R.H. Waldemer, P.G. Tratnyek, R.L. Johnson, J.T. Nurmi, Oxidation of chlorinated ethenes by heat-activated persulfate: Kinetics and products, Environ. Sci. Technol. 41 (2007) 1010–1015. https://doi.org/10.1021/es062237m.

[56]   G. Moussavi, M. Pourakbar, E. Aghayani, M. Mahdavianpour, S. Shekoohiyan, Comparing the efficacy of VUV and UVC/$S_2O_8^{2-}$ advanced oxidation processes for degradation and mineralization of cyanide in wastewater, Chem. Eng. J. 294 (2016) 273–280. https://doi.org/10.1016/j.cej.2016.02.113.

[57]   T. Olmez-Hanci, I. Arslan-Alaton, Comparison of sulfate and hydroxyl radical based advanced oxidation of phenol, Chem. Eng. J. 224 (2013) 10–16. https://doi.org/10.1016/j.cej.2012.11.007.

[58]   B.T. Oh, Y.S. Seo, D. Sudhakar, J.H. Choe, S.M. Lee, Y.J. Park, M. Cho, Oxidative degradation of endotoxin by advanced oxidation process (O3/H2O2 & UV/H2O2), J. Hazard. Mater. 279 (2014) 105–110. https://doi.org/10.1016/j.jhazmat.2014.06.065.

[59]   L.W. Gassie, J.D. Englehardt, Mineralization of greywater organics by the ozone-UV advanced oxidation process: Kinetic modeling and efficiency, Environ. Sci. Water Res. Technol. 5 (2019) 1956–1970. https://doi.org/10.1039/c9ew00653b.

[60]   L.A. Gonzalez-Burciaga, C.M. Nunez-Nunez, M.M. Morones-esquivel, M. Avila-Santos, A. Lemus-Santana, J. B. Proal-Najera, Characterization and Comparative Performance of TiO_2 Photocatalysts on 6-Mercaptopurine Degradation by Solar Heterogeneous Photocatalysis, Catalysts. 10 (2020) 118. https://doi.org/10.3390/catal10010118

[61]   M. Pirsaheb, H. Hossaini, H. Janjani, Reclamation of hospital secondary treatment effluent by sulfate radicals based–advanced oxidation processes (SR-AOPs) for removal of antibiotics, Microchem. J. 153 (2020) 104430. https://doi.org/10.1016/j.microc.2019.104430.

[62]   A. Hossein Panahi, A. Meshkinian, S.D. Ashrafi, M. Khan, A. Naghizadeh, G. Abi, H. Kamani, Survey of sono-activated persulfate process for treatment of real dairy wastewater, Int. J. Environ. Sci. Technol. 17 (2020) 93–98. https://doi.org/10.1007/s13762-019-02324-4.

[63]   A. Shokri, K. Mahanpoor, D. Soodbar, Degradation of Ortho-Toluidine in petrochemical wastewater by ozonation, UV/O_3, O_3/H_2O_2 and $UV/O_3/H_2O_2$ processes, Desalin. Water Treat. 57 (2016) 16473–16482. https://doi.org/10.1080/19443994.2015.1085454.

[64]   M. Sayed, J.A. Khan, L.A. Shah, N.S. Shah, H.M. Khan, F. Rehman, A.R. Khan, A.M. Khan, Degradation of quinolone antibiotic, norfloxacin, in aqueous solution using gamma-ray irradiation, Environ. Sci. Pollut. Res. 23 (2016) 13155–13168. https://doi.org/10.1007/s11356-016-6475-x.

[65]   J. Li, S. Wang, Y. Li, L. Wang, T. Xu, Y. Zhang, Z. Jiang, Supercritical water oxidation of semi-coke wastewater: Effects of operating parameters, reaction mechanism and process enhancement, Sci. Total Environ. 710 (2020) 134396. https://doi.org/10.1016/j.scitotenv.2019.134396.

[66]   Q. Jing, H. Li, Hierarchical nickel cobalt oxide spinel microspheres catalyze mineralization of humic substances during wet air oxidation at atmospheric pressure, Appl. Catal. B Environ. 256 (2019) 117858. https://doi.org/10.1016/j.apcatb.2019.117858.

[67]   J.R. Domínguez, M.J. Muñoz, P. Palo, T. González, J.A. Peres, E.M. Cuerda-Correa, Fenton advanced oxidation of emerging pollutants: Parabens, Int. J. Energy Environ. Eng. 5 (2014) 89. https://doi.org/10.1007/s40095-014-0089-1.

[68]   M.S. Yahya, N. Oturan, K. El Kacemi, M. El Karbane, C.T. Aravindakumar, M.A. Oturan, Oxidative degradation study on antimicrobial agent ciprofloxacin by electro-

fenton process: Kinetics and oxidation products, Chemosphere. 117 (2014) 447–454. https://doi.org/10.1016/j.chemosphere.2014.08.016.

[69]   J.H. Kim, H.K. Lee, Y.J. Park, S.B. Lee, S.J. Choi, W. Oh, H.S. Kim, C.R. Kim, K.C. Kim, B.C. Seo, Studies on decomposition behavior of oxalic acid waste by UVC photo-Fenton advanced oxidation process, Nucl. Eng. Technol. 51 (2019) 1957–1963. https://doi.org/10.1016/j.net.2019.06.011.

[70]   K. Sivagami, D. Anand, G. Divyapriya, I. Nambi, Treatment of petroleum oil spill sludge using the combined ultrasound and Fenton oxidation process, Ultrason. Sonochem. 51 (2019) 340–349. https://doi.org/10.1016/j.ultsonch.2018.09.007.

[71]   N. Remya, A. Swain, Soft drink industry wastewater treatment in microwave photocatalytic system – Exploration of removal efficiency and degradation mechanism, Sep. Purif. Technol. 210 (2019) 600–607. https://doi.org/10.1016/j.seppur.2018.08.051.

[72]   S. Zuo, D. Li, H. Xu, D. Xia, An integrated microwave-ultraviolet catalysis process of four peroxides for wastewater treatment: Free radical generation rate and mechanism, Chem. Eng. J. 380 (2020) 122434. https://doi.org/10.1016/j.cej.2019.122434.

[73]   A.L. Garcia-Costa, L. Lopez-Perela, X. Xu, J.A. Zazo, J.J. Rodriguez, J.A. Casas, Activated carbon as catalyst for microwave-assisted wet peroxide oxidation of aromatic hydrocarbons, Environ. Sci. Pollut. Res. 25 (2018) 27748–27755. https://doi.org/10.1007/s11356-018-2291-9.

[74]   H. Milh, K. Van Eyck, R. Dewil, Degradation of 4-chlorophenol by microwave-enhanced advanced oxidation processes: Kinetics and influential process parameters, Water (Switzerland). 10 (2018) 247. https://doi.org/10.3390/w10030247.

[75]   Y. Zhang, Y. Hou, F. Chen, Z. Xiao, J. Zhang, X. Hu, The degradation of chlorpyrifos and diazinon in aqueous solution by ultrasonic irradiation: Effect of parameters and degradation pathway, Chemosphere. 82 (2011) 1109–1115. https://doi.org/10.1016/j.chemosphere.2010.11.081.

[76]   E. Villaroel, J. Silva-Agredo, C. Petrier, G. Taborda, R.A. Torres-Palma, Ultrasonic degradation of acetaminophen in water: Effect of sonochemical parameters and water matrix, Ultrason. Sonochem. 21 (2014) 1763–1769. https://doi.org/10.1016/j.ultsonch.2014.04.002.

[77]   S. Papoutsakis, S. Miralles-Cuevas, N. Gondrexon, S. Baup, S. Malato, C. Pulgarin, Coupling between high-frequency ultrasound and solar photo-Fenton at pilot

scale for the treatment of organic contaminants: An initial approach, Ultrason. Sonochem. 22 (2015) 527–534. https://doi.org/10.1016/j.ultsonch.2014.05.003.

[78]   X. Hu, X. Wang, Y. Ban, B. Ren, A comparative study of UV-Fenton, UV-H_2O_2 and Fenton reaction treatment of landfill leachate, Environ. Technol. 32 (2011) 945–951. https://doi.org/10.1080/09593330.2010.521953.

[79]   J. Tejera, R. Miranda, D. Hermosilla, I. Urra, C. Negro, Á. Blanco, Treatment of a mature landfill leachate: Comparison between homogeneous and heterogeneous photo-fenton with different pretreatments, Water (Switzerland). 11 (2019) 1849. https://doi.org/10.3390/w11091849.

[80]   J. Beltran-Heredia, J. Torregrosa, J.R. Dominguez, J.A. Peres, Comparison of the degradation of p-hydroxybenzoic acid in aqueous solution by several oxidation processes, Chemosphere. 42 (2001) 351–359. https://doi.org/10.1016/S0045-6535(00)00136-3.

[81]   S. Esplugas, J. Giménez, S. Contreras, E. Pascual, M. Rodríguez, Comparison of different advanced oxidation processes for phenol degradation, Water Res. 36 (2002) 1034–1042. https://doi.org/10.1016/S0043-1354(01)00301-3.

[82]   M. Muruganandham, S.H. Chen, J.J. Wu, Mineralization of N-methyl-2-pyrolidone by advanced oxidation processes, Sep. Purif. Technol. 55 (2007) 360–367. https://doi.org/10.1016/j.seppur.2007.01.009.

[83]   M.R. Assalin, S.G. de Moraes, S.C.N. Queiroz, V.L. Ferracini, N. Duran, Studies on degradation of glyphosate by several oxidative chemical processes: Ozonation, photolysis and heterogeneous photocatalysis, J. Environ. Sci. Heal. - Part B Pestic. Food Contam. Agric. Wastes. 45 (2010) 89–94. https://doi.org/10.1080/03601230903404598.

[84]   K. Kovács, J. Farkas, G. Veréb, E. Arany, G. Simon, K. Schrantz, A. Dombi, K. Hernádi, T. Alapi, Comparison of various advanced oxidation processes for the degradation of phenylurea herbicides, J. Environ. Sci. Heal. - Part B Pestic. Food Contam. Agric. Wastes. 51 (2016) 205–214. https://doi.org/10.1080/03601234.2015.1120597.

[85]   G. Boczkaj, A. Fernandes, P. Makoś, Study of Different Advanced Oxidation Processes for Wastewater Treatment from Petroleum Bitumen Production at Basic pH, Ind. Eng. Chem. Res. 56 (2017) 8806–8814. https://doi.org/10.1021/acs.iecr.7b01507.

[86]   S.B. Kausley, K.S. Desai, S. Shrivastava, P.R. Shah, B.R. Patil, A.B. Pandit, Mineralization of alkyd resin wastewater: Feasibility of different advanced oxidation

processes, J. Environ. Chem. Eng. 6 (2018) 3690–3701.
https://doi.org/10.1016/j.jece.2017.04.001.

[87]   M. Muruganandham, K. Selvam, M. Swaminathan, A comparative study of
quantum yield and electrical energy per order (E_{Eo}) for advanced oxidative
decolourisation of reactive azo dyes by UV light, J. Hazard. Mater. 144 (2007) 316–
322. https://doi.org/10.1016/j.jhazmat.2006.10.035.

[88]   Y. Laftani, A. Boussaoud, B. Chatib, M. Hachkar, M. El Makhfouk, M. Khayar,
Comparison of advanced oxidation processes for degrading ponceau S dye:
Application of the photo-fenton process, Maced. J. Chem. Chem. Eng. 38 (2019) 197–
205. https://doi.org/10.20450/mjcce.2019.1888.

[89]   K. Kowalska, G. Maniakova, M. Carotenuto, O. Sacco, V. Vaiano, G. Lofrano, L.
Rizzo, Removal of carbamazepine, diclofenac and trimethoprim by solar driven
advanced oxidation processes in a compound triangular collector based reactor: A
comparison between homogeneous and heterogeneous processes, Chemosphere. 238
(2020) 124665. https://doi.org/10.1016/j.chemosphere.2019.124665.

[90]   G. Maniakova, K. Kowalska, S. Murgolo, G. Mascolo, G. Libralato, G. Lofrano,
O. Sacco, M. Guida, L. Rizzo, Comparison between heterogeneous and homogeneous
solar driven advanced oxidation processes for urban wastewater treatment:
Pharmaceuticals removal and toxicity, Sep. Purif. Technol. 236 (2020) 116249.
https://doi.org/10.1016/j.seppur.2019.116249.

[91]   J.F. Pérez, J. Llanos, C. Sáez, C. López, P. Cañizares, M.A. Rodrigo, Treatment of
real effluents from the pharmaceutical industry: A comparison between Fenton
oxidation and conductive-diamond electro-oxidation, J. Environ. Manage. 195 (2017)
216–223. https://doi.org/10.1016/j.jenvman.2016.08.009.

[92]   Information on http://labees.civil.fau.edu/Final_report_Englehardt.pdf

[93]   S. Krishnan, H. Rawindran, C.M. Sinnathambi, J.W. Lim, Comparison of various
advanced oxidation processes used in remediation of industrial wastewater laden with
recalcitrant pollutants, IOP Conf. Ser. Mater. Sci. Eng. 206 (2017) 012089.
https://doi.org/10.1088/1757-899X/206/1/012089.

[94]   H. Barndõk, D. Hermosilla, C. Negro, Á. Blanco, Comparison and Predesign Cost
Assessment of Different Advanced Oxidation Processes for the Treatment of 1,4-
Dioxane-Containing Wastewater from the Chemical Industry, ACS Sustain. Chem.
Eng. 6 (2018) 5888–5894. https://doi.org/10.1021/acssuschemeng.7b04234.

[95]   P. Asaithambi, R. Saravanathamizhan, M. Matheswaran, Comparison of treatment

and energy efficiency of advanced oxidation processes for the distillery wastewater, Int. J. Environ. Sci. Technol. 12 (2015) 2213–2220. https://doi.org/10.1007/s13762-014-0589-9.

[96] E.J. Rosenfeldt, K.G. Linden, S. Canonica, U. von Gunten, Comparison of the efficiency of OH radical formation during ozonation and the advanced oxidation processes O_3/H_2O_2 and UV/H_2O_2, Water Res. 40 (2006) 3695–3704. https://doi.org/10.1016/j.watres.2006.09.008.

[97] K. Paździor, L. Bilińska, S. Ledakowicz, A review of the existing and emerging technologies in the combination of AOPs and biological processes in industrial textile wastewater treatment, Chem. Eng. J. 376 (2019) 120597. https://doi.org/10.1016/j.cej.2018.12.057.

[98] J. Wang and L. Xu, AOPs for municipal and industrial wastewater treatment, in: M.I. Stefan (Ed.), Advanced Oxidation Processes for Water Treatment, IWA Publishing, 2018, pp.631-666.

[99] Stuart J. Khan, Troy Walker, Benjamin D. Stanford and Jörg E. Drewes, Advanced treatment for potable water reuse, in: M.I. Stefan (Ed.), Advanced Oxidation Processes for Water Treatment, IWA Publishing, 2018,pp. 581-606.

[100] C. Comninellis, A. Kapalka, S. Malato, S. A. Parsons, I. Poulios and D. Mantzavinos, Advanced oxidation processes for water treatment: advances and trends for R&D, J. Chem. Technol. Biotechnol. 83 (2008) 1163–1169. https://doi.org/10.1002/jctb.

[101] L. Patria, C. Maugans, C. Ellis, M. Belkhodja, D. Cretenot, F. Luck and B. Copa, Wet air oxidation processes,   in: S. Parsons (Ed.), Advanced oxidation processes for water and wastewater treatment, IWA publishers, 2018, pp. 247-274. https://doi.org/10.1007/s11356-018-3411-2.

[102] T.J. Mason and C. Pétrier, Ultrasound processes, in: S. Parsons (Ed.), Advanced oxidation processes for water and wastewater treatment, IWA publishers, 2018, pp.185-208. https://doi.org/10.1007/s11356-018-3411-2

[103] S. Malato, J. Blanco, D.C. Alarcón, M.I. Maldonado, P. Fernández-Ibáñez, W. Gernjak, Photocatalytic decontamination and disinfection of water with solar collectors, Catal. Today. 122 (2007) 137–149. https://doi.org/10.1016/j.cattod.2007.01.034.

[104] J. Wang, R. Zhuan, L. Chu, The occurrence, distribution and degradation of antibiotics by ionizing radiation: An overview, Sci. Total Environ. 646 (2019) 1385–

1397. https://doi.org/10.1016/j.scitotenv.2018.07.415.

[105] B. Garza-Campos, E. Brillas, A. Hernández-Ramírez, A. El-Ghenymy, J.L. Guzmán-Mar, E.J. Ruiz-Ruiz, Salicylic acid degradation by advanced oxidation processes. Coupling of solar photoelectro-Fenton and solar heterogeneous photocatalysis, J. Hazard. Mater. 319 (2016) 34–42. https://doi.org/10.1016/j.jhazmat.2016.02.050.

[106] H. Zazou, H. Afanga, S. Akhouairi, H. Ouchtak, A.A. Addi, R.A. Akbour, A. Assabbane, J. Douch, A. Elmchaouri, J. Duplay, A. Jada, M. Hamdani, Treatment of textile industry wastewater by electrocoagulation coupled with electrochemical advanced oxidation process, J. Water Process Eng. 28 (2019) 214–221. https://doi.org/10.1016/j.jwpe.2019.02.006.

[107] D. Cassano, A. Zapata, G. Brunetti, G. Del Moro, C. Di Iaconi, I. Oller, S. Malato, G. Mascolo, Comparison of several combined/integrated biological-AOPs setups for the treatment of municipal landfill leachate: Minimization of operating costs and effluent toxicity, Chem. Eng. J. 172 (2011) 250–257. https://doi.org/10.1016/j.cej.2011.05.098.

[108] H. Liu, L. Fang, J. Wang, C. Lin, Supercritical water oxidation of p-aminonaphthalenesulfonic acid and enhancement by Fenton agent, Environ. Technol. (United Kingdom). (2019) 1–23. https://doi.org/10.1080/09593330.2019.1571115.

[109] O. Ganzenko, C. Trellu, S. Papirio, N. Oturan, D. Huguenot, E.D. van Hullebusch, G. Esposito, M.A. Oturan, Bioelectro-Fenton: evaluation of a combined biological—advanced oxidation treatment for pharmaceutical wastewater, Environ. Sci. Pollut. Res. 25 (2018) 20283–20292. https://doi.org/10.1007/s11356-017-8450-6.

[110] T.K. Nivya, T. Minimol Pieus, Comparison of Photo ElectroFenton Process(PEF) and combination of PEF Process and Membrane Bioreactor in the treatment of Landfill Leachate, Procedia Technol. 24 (2016) 224–231. https://doi.org/10.1016/j.protcy.2016.05.030.

[111] M. Sun, Y. Zhang, S.Y. Kong, L.F. Zhai, S. Wang, Excellent performance of electro-assisted catalytic wet air oxidation of refractory organic pollutants, Water Res. 158 (2019) 313–321. https://doi.org/10.1016/j.watres.2019.04.040.

[112] R. Poblete, E. Cortes, G. Salihoglu, N.K. Salihoglu, Ultrasound and heterogeneous photocatalysis for the treatment of vinasse from pisco production, Ultrason. Sonochem. 61 (2020) 104825. https://doi.org/10.1016/j.ultsonch.2019.104825.

[113] R. Bonora, C. Boaretti, L. Campea, M. Roso, A. Martucci, M. Modesti, A.

Lorenzetti, Combined AOPs for formaldehyde degradation using heterogeneous nanostructured catalysts, Nanomaterials. 10 (2020) 148. https://doi.org/10.3390/nano10010148.

[114] X. Zheng, Z.-P. Shen, L. Shi, R. Cheng, D.-H. Yuan, Photocatalytic Membrane Reactors (PMRs) in Water Treatment: Configurations and Influencing Factors, Catalysts. 7 (2017) 224. https://doi.org/10.3390/catal7080224.

[115] J. Gómez-pastora, S. Dominguez, E. Bringas, M.J. Rivero, I. Ortiz, D.D. Dionysiou, Review and perspectives on the use of magnetic nanophotocatalysts (MNPCs) in water treatment, Chem. Eng. J. 310 (2017) 407–427. https://doi.org/10.1016/j.cej.2016.04.140.

[116] T. Leshuk, H. Krishnakumar, D. de O. Livera, F. Gu, Floating photocatalysts for passive solar degradation of naphthenic acids in oil sands process-affected water, Water (Switzerland). 10 (2018) 202. https://doi.org/10.3390/w10020202.

[117] T.S. Rajaraman, S.P. Parikh, V.G. Gandhi, Black TiO2: A review of its properties and conflicting trends, Chem. Eng. J. 389 (2020) 123918. https://doi.org/10.1016/j.cej.2019.123918.

[118] J. Rodríguez-Chueca, M.P. Ormad, R. Mosteo, J. Sarasa, J.L. Ovelleiro, Conventional and Advanced Oxidation Processes Used in Disinfection of Treated Urban Wastewater, Water Environ. Res. 87 (2015) 281–288. https://doi.org/10.2175/106143014x13987223590362.

[119] Y. Deng, Advanced oxidation processes (AOPs) for reduction of organic pollutants in landfill leachate: A review, Int. J. Environ. Waste Manag. 4 (2009) 366–384. https://doi.org/10.1504/IJEWM.2009.027402.

Chapter 3

Degradation of Pharmaceutical Pollutants under UV Light using TiO₂ Nanomaterial Synthesized through Reverse Micelle Nanodomains

K.S. Varma[1], B. Bharatiya[3], R.J. Tayade[2], A.D. Shukla[1], P.A. Joshi[1], Vimal Gandhi[1]*

[1]Department of Chemical Engineering & Shah Schulman Center for Surface Science and Nanotechnology, Dharmsinh Desai University, College Road, Nadiad 387 001, Gujarat, India

[2]Discipline of Inorganic Materials and Catalysis, Central Salt and Marine Chemicals Research Institute, Council of Scientific and Industrial Research (CSIR), G. B. Marg, Bhavnagar-364002, Gujarat, India

[3]School of Chemistry, University of Bristol, England

* vggandhi.ch@ddu.ac.in

Abstract

Controlling water pollution are huge challenges throughout the world especially concerning pharmaceutical pollutants. Common practices at industrial wastewater treatment facilities need to be upgraded with advanced wastewater treatment techniques. TiO₂ based photocatalytic processes have shown great potential for removal of these aqueous pharmaceutical pollutants. Reverse micelle based modified sol-gel method is utilized for the synthesis of TiO₂ nanomaterial. Generated reverse micelle nanodomains have controlled size and particle size distribution (PSD) of synthesized TiO₂ nanomaterial, as revealed by SEM and DLS analysis. Thermal behaviour of synthesized sample is characterized by TGA analysis. TiO₂ photocatalyst is also characterized through XRD, BET surface area, and UV-Vis spectroscopy. TiO₂ photocatalyst is used for degradation of three model pharmaceutical pollutants viz. Levofloxacin hemihydrate (LFX), Metronidazole (MNZ) and Ketorolac tromethamine (KRL) under a UV light source. Reverse micelle mediated modified sol-gel method synthesized TiO₂ nanomaterial has shown excellent photocatalytical performance, where degradation efficiency of LFX, KRL and MNZ were found to be 99.6%, 98% and 91.4% respectively within a little as 60 minutes.

Keywords

Pharmaceutical Pollutants, Degradation, Photocatalyst, TiO₂ Nanomaterial, Reverse Micelle, UV Light

Advances in Wastewater Treatment I
Materials Research Foundations **91** (2021) 87-110

Materials Research Forum LLC
https://doi.org/10.21741/9781644901144-3

Contents

1. Introduction

The world economy is hugely depending upon industrial development along with urbanization. The pharmaceutical sector is a rapidly expanding industrial sector, due to substantial demand of antibiotics, nonsteroidal anti-inflammatory drugs (NSAIDs) and many other products. It is found that environmental issues related to antibiotic emissions are higher throughout the world especially in Asian countries [1]. Generation of wastewater through pharmaceutical industries are contaminating the water matrices, which triggering several adverse effects on human and aquatic ecosystems. Complete removal of these pharmaceutical pollutants through conventional wastewater treatment techniques are huge challenges in front of research and industrial communities due to their persistent characteristics [2,3].

These industrial circumstances demand that advanced wastewater treatment techniques must be developed and incorporated alongside with conventional wastewater treatment techniques [4,5]. Membrane filtration and advanced oxidation processes (AOPs) are widely employed as advanced wastewater treatment techniques [6]. For treating diverse complex organic pollutants, AOPs is much explored in recent times due to its ability in complete removal of pollutants using generated hydroxyl ($OH^\bullet$) and superoxide ($O_2^{-\bullet}$) radicals [7]. There are different types of AOPs such as Fenton, photo-fenton [8,9], ozone-based processes [10], photolysis, photoelectrocatalysis and photocatalysis [11,12], which are utilized for removing the pollutants from contaminated water. From different AOPs techniques, one of the most promising techniques is photocatalysis, which is shown effective degradation of various pollutants and exhibited notable environmental application [13-16]. In photocatalysis, catalyst material (mostly semiconductor nanomaterials) is activated under appropriate light energy and initiated photoreaction, without being consumed during the reaction [17].

Many types of photocatalytic materials (TiO_2, ZnO, ZnS, MOS_2, Fe_2O_3, WO_3 and perovskites) are explored for eliminating the contaminants from polluted water. Amongst these materials, titanium dioxide (TiO_2) is widely acceptable by research communities due to its nontoxicity, stability and their higher strength in oxidization of pollutants [18,19]. For synthesis of TiO_2 and other photocatalytic nanomaterials, different methods which are commonly used are sol-gel, solvothermal, hydrothermal and microemulsion [20-25]. One of the most economical and simple synthesis methods for nanomaterials is

Advances in Wastewater Treatment I   Materials Research Forum LLC
Materials Research Foundations **91** (2021) 87-110   https://doi.org/10.21741/9781644901144-3

sol-gel. However, effective control over the nucleation process is the key limitation of the sol-gel method, which resulted in the wide size distribution of nanomaterials. Reverse micelle based modified sol-gel method is providing better control over the particle size distribution (PSD) and shape of the nanomaterials as hydrolysis reaction is governed by generated reverse micelles [26,27].

For complex organic pharmaceutical pollutants, TiO_2 nanomaterials is exhibited significant degradation performance due to its high oxidization ability under the UV light source. TiO_2 nanomaterials is utilized for degradation of sulfamethoxazole drug under UV-A light source, which has shown more than 95% drug compound degradation in 60 minutes only [28]. TiO_2 nanomaterials suspension is effectively degraded oxolinic acid and also provided support to reduce the toxicity of antibiotic solution by 60% [29]. Under optimized condition of pH, initial pollutant concentration and catalyst loading, complete degradation of benzylparaben was occurred by using TiO_2 during the 120 minutes UV light irradiation [30]. UV light mediated degradation performance was checked against tylosin by using TiO_2 nanomaterials, shown the good photocatalytic performance against this antibiotic pollutant [31].

In the present study, TiO_2 nanomaterial is synthesized through reverse micelle mediated modified sol-gel method. UV light mediated photocatalytic degradation of three pharmaceutical pollutants LFX, KRL and MNZ are carried out using TiO_2 nanomaterial. Moreover, effect of catalyst loading and initial pollutant concentration on LFX degradation along with its kinetics are discussed here.

2.   Mechanism of TiO_2 photocatalyst against organic pollutants under UV light source

Under the light source, semiconductor TiO_2 material is functioned as photocatalyst when the photon energy exceeded the band gap energy of TiO_2 which is around 3.2 eV (for anatase phase). Mechanism of TiO_2 against organic pollutant under UV light is illustrated as in Fig. 1. Photocatalytic reaction is initiated as electron (e^-) is excited from valence band (VB) to conduction band (CB), leads to creation of electron-free void which is specified as hole (h^+) and recombination of electron-hole pair is immediately occurred at swift rate (Eq. 1 and 2).

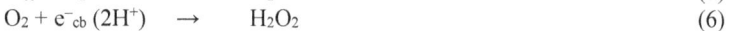

$$TiO_2 + h\nu \quad \rightarrow \quad e^-_{cb} + h^+_{vb} \qquad (1)$$
$$h^+_{vb} + e^-_{cb} \quad \rightarrow \quad heat \qquad (2)$$
$$h^+_{vb} + OH^-_{ads} \quad \rightarrow \quad OH^{\bullet}_{ads} \qquad (3)$$
$$h^+_{vb} + H_2O \quad \rightarrow \quad OH^{\bullet} + H^+ \qquad (4)$$
$$e^-_{cb} + O_2 \quad \rightarrow \quad O_2^{-\bullet} \qquad (5)$$
$$O_2 + e^-_{cb} (2H^+) \quad \rightarrow \quad H_2O_2 \qquad (6)$$

$$H_2O_2 + O_2^{-\bullet} \quad \rightarrow \quad OH^{\bullet}_{ads} + OH^- + O_2 \quad\quad (7)$$

$$H_2O_2 + e^-_{cb} \quad \rightarrow \quad OH^{\bullet}_{ads} + OH^- \quad\quad (8)$$

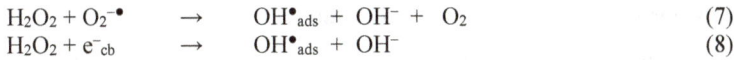

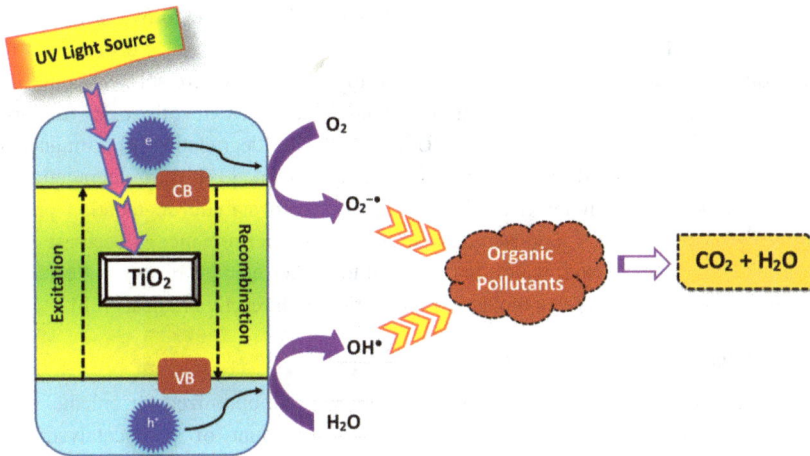

Figure 1. Mechanism of TiO₂ photocatalyst against organic pollutants under UV light source.

At the surface of TiO_2, adsorbed hydroxyl group (OH^-_{ads}) is formed due to interaction of holes and surface adsorbed water or hydroxyl group (Eq. 3). Migration of surface adsorbed hydroxyl radical ($OH^{\bullet}_{ads}$) to bulk solution, resulted in the formation of highly reactive free OH radical ($OH^{\bullet}$) as stated in Eq. 4. At the conduction band, superoxide radical ($O_2^{-\bullet}$) due to interaction of oxygen with free electron (Eq. 5). H_2O_2 is formed and their interaction with superoxide radical and free electron is illustrated in Eq. 6, 7 and 8. Due to highly reactive free radicals, organic pollutants are degraded in presence of light energy and are converted to CO_2, H_2O and other intermediate products [32,33].

3.   Factors affecting the photocatalyst performance

Performance of heterogenous photocatalytic process is influenced by many processing parameters. Key factors which are influencing the performance of the photocatalytic process are initial concentration of pollutant, catalyst loading, pH and light intensity. Effect of different parameters on degradation performance are extensively reviewed for diverse aqueous pollutants such as pharmaceuticals, dyes and many others [34,35].

Advances in Wastewater Treatment I   Materials Research Forum LLC
Materials Research Foundations **91** (2021) 87-110   https://doi.org/10.21741/9781644901144-3

3.1   Influence of initial concentration of pollutant

It is important to choose appropriate concentration of pollutant as photocatalytic process is occurred on the surface of the photocatalyst. There is the possibility of the declining in the degradation performance at high concentration of pollutant, due to over saturation of reactive sites [36]. It is found that degradation rate of phthalic acid is decreased when initial concentration is increased from 100 mg/L to 300 mg/L, where sol-gel synthesized TiO$_2$ nanomaterials is used [20]. It is found that when levofloxacin antibiotic concentration is increases from 50 mg/L to 75 mg/L, degradation performance of photocatalyst is rapidly decreases from 95.4% to 30.7% [37]. For degradation of methylene blue, TiO$_2$ nanomaterials is utilized for photocatalytic degradation process under UV-A light source. It is observed that degradation performance is decreases with increasing dye concentration, which may be due to higher amount of dye molecules on TiO$_2$ surface are interrupted dye molecules interaction with reactive radicals [38].

3.2   Influence of catalyst loading

Concentration of pollutant inside the photoreactor must be optimized for enhancing the efficiency of the photocatalytic process. At high amount of photocatalyst, light penetrability to the surface of catalysts are decreased, which affected the degradation efficiency [36]. Degradation of tetracycline antibiotic under UV light is carried out using TiO$_2$ nanomaterials. It is found that at optimum TiO$_2$ photocatalyst loading of 1g/L, highest antibiotic degradation is occurred. Photocatalytic degradation performance is decreased at higher photocatalyst loading which may be due to inadequate light (photon) interactions with nanomaterials or aggregation of nanomaterials could occurred [39]. Effect of various TiO$_2$ photocatalyst loading on degradation performance against 100 mg/l phthalic acid is checked, where it is observed that optimum catalyst concentration is 2.5 g/L. Degradation efficiency is decreased with further increment in catalyst loading which may influenced the photons interaction with nanomaterials and pollutant solution [20]. Ciprofloxacin antibiotic is degraded using TiO$_2$ nanomaterials supported on montmorillonite under UV light. It is found that degradation efficiency is increases when catalyst loading is increased from 0.025 g/L to 0.1 g/L and at higher catalyst loading of 0.15 g/L, degradation performance is decreases which may be due to reduction in reactive sites resulted from aggregation of nanomaterials [40]. Degussa P25 nanomaterials is utilized for degradation of three different pharmaceutical aqueous pollutant solution i.e., sulfisoxazole, sulfachlorpyridazine and sulfapyridine under UV light irradiation. They are observed that degradation performance is increased with increment of catalyst amount for all targeted pollutants [41].

3.3   Influence of pH

The change in pH may altered the degree of ionization, surface charge of catalyst and also adsorption capacity of pollutant molecules on the surface of the photocatalyst [36]. Degradation of ciprofloxacin antibiotic aqueous solution under UV-C light source is carried out using commercial TiO_2 nanomaterials. It is found that degradation efficiency is enhanced greatly when pH is increased from 3 to 5. However, further increment in pH is affecting the degradation performance, which may be due to electrostatic repulsion between TiO_2 and negatively charged antibiotic molecules [42]. Effect of pH on phenol degradation is checked using immobilized TiO_2 type photoreactor system. It is observed that removal efficiency of phenol from aqueous pollutant solution is decreased when pH is changed from acidic range (pH of 3) to alkaline (pH of 11) [43]. Five different aqueous pollutants viz. gemfibrozil, n-nitrosodimethylamine, 1,4 dioxane, tris-2-chloroethyl phosphate and 17β estradiol are targeted for photocatalytic degradation application in the presence of TiO_2 nanomaterials. Effect of pH on degradation of pollutants are checked, where it is observed that at pH of 5, highest degradation is occurred. However, at pH of 3 and pH of 9, degradation performance is declined for all five pollutants, which may be due to alternation of surface charge of catalyst [44]. Influence of pH (4-10) is checked on degradation efficiency of three dye pollutants Reactive Red 195, Reactive Black 5 and Alizarin Cyanine Green G. It is found that the highest degradation is observed in acidic pH of 4 for all targeted aqueous pollutants due to interactions of dye molecules and positively charged catalyst surface through electrostatic force. While, performance of photocatalyst is declined at alkaline pH of 10 [45]. Influence of pH on degradation efficiency for different types of pollutants are distinct due to their individual interaction behavior with the catalyst. It is found that degradation performance in Acid Orange 7 dye is much affected compared to Alizarin Cyanine Green dye pollutant, when pH is varied from 2 to 10 under similar catalyst loading and pollutant concentration [46].

3.4   Influence of light intensity

The degradation efficiency is primarily depending upon the number of generated electron-hole pairs through light energy. Hence, light intensity is also a crucial factor for the photocatalytic process [36]. Degradation performance are compared for five commercial TiO_2 based nanomaterials against carbamazepine under UV light source. P90 photocatalyst showed best photocatalytical activity compared to others. Effect of light intensity on degradation of pharmaceutical aqueous pollutant is checked, where degradation efficiency is improved when light intensity is increased from 15.08 W/m^2 to 18.94 W/m^2. At 24.2 W/m^2 light intensity, photocatalytical activity is declined, because degradation rate is not much relied upon radiant flux at higher light intensity [47].

Change in light intensity can altered the photocatalytical activity. It is found that degradation efficiency of TiO_2 nanomaterial against furfural is significantly decreased from 93% to 50%, when UV light intensity is altered from 125W to 30W [48].

4. Experiment

4.1 Chemicals & materials

Titanium tetraisopropxide (TTIP) (>98%) was procured from Spectrochem Pvt. Ltd., Mumbai. Hexane (fraction from petroleum) was procured from S D Fine-Chem Limited, Mumbai. Triton X-114 (laboratory grade) was procured from Sigma-Aldrich Chemicals Pvt. Ltd., Bangalore. Toluene (≥ 99.0 %) was procured from Merck Speciallities Pvt. Ltd., Mumbai. Degussa P25 TiO_2 catalyst was given by Degussa. For synthesis, characterization and photoactivity of TiO_2 material, millipore water was utilized. LFX, MNZ, KRL were gifted by Rhombus pharma private limited, Aarti Drugs Ltd., and Surya Life Sciences Ltd., Gujrat, India respectively.

4.2 Synthesis of TiO_2 photocatalyst

TiO_2 nanomaterials was synthesized through reverse micelle mediated modified sol-gel method. Addition of solvent mixture consists of 210 ml hexane and 90 ml toluene into 15.87 gm of triton X-114 non-ionic surfactant in a conical flask resulted the formation of reverse micelle, which was kept under stirring condition for around 1 H. Millipore water of 900 µl was added drop by drop into surfactant-solvent mixture under stirring state for about 1 H. Reverse micelle was bloated due to inclusion of water micro-droplets inside their core, which are termed as hydrated reverse micelle. TTIP is added at very slow rate into hydrated reverse micelle system under constant mixing condition. Hydrolysis of TTIP was occurred inside aqueous core of reverse micelle, where transparent sol was turned into viscous milky white solution. The sol mixture was kept under stirring condition for around 15 H. The resulted gel mixture was centrifuged at 8000 rpm for 10 min. in ultracentrifuge for separating the gel and solvent-surfactant mixture. The gel was left in petridis exposed to air for around 6 H. Drying of gel was carried out in oven at 130°C for 18 hrs. The dried powder was calcined at 400°C for 4 hrs in a muffle furnace, where temperature increment rate is around 10°C/m

4.3 Characterization of photocatalyst

Thermogravimetric analysis (TGA) was performed for exploring thermal characteristics of dried $Ti(OH)_4$ sample using Mettler Toledo TGA/DSC thermal analyser in the temperature range from 25 to 700°C where rate of heating is 10 °C/min.

Advances in Wastewater Treatment I
Materials Research Foundations **91** (2021) 87-110

Materials Research Forum LLC
https://doi.org/10.21741/9781644901144-3

For phase identification and measurement of crystallite size, XRD experiment was performed using Bruker AXS (D8 advance) X-ray powder diffractometer with Cu Kα radiation (λ = 1.5418 Å), where scanning of TiO$_2$ sample was in 2θ range of 10-85°. Rutile phase content in nanomaterials are calculated by using Spurr equation as shown in Eq. 9 [49].

$$F_R = \frac{1}{1+0.8[I_A(101)/I_R(110)]} \qquad (9)$$

Where $I_A(101)$ and $I_R(110)$ are anatase and rutile peak respectively and F_R is quantity of rutile phase in fraction.

Average crystallite size is estimated using Scherer equation as shown in Eq. 10 [50].

$$D = \frac{0.94\lambda}{\beta\cos\theta} \qquad (10)$$

Where D is average crystallite size of nanomaterial, λ = 1.5406 Å, θ is the Bragg angle and β is full width at half maximum intensity peak.

To find the size and shape of the TiO$_2$ nanomaterials, scanning electron microscope (SEM) of JEOL, JSM-7600 F model was utilized with operating voltage of 5-10 kV.

The average particle size and particle size distribution (PSD) of was analysed using dynamic light scattering (DLS) using Malvern particle size analyser, where TiO$_2$ nanomaterials were dispersed in Millipore water and sonicated for around 30 min to isolate the agglomerated particles.

Brunauer–Emmett–Teller (BET) surface area and allied properties of TiO$_2$ nanomaterials is investigated through N$_2$ gas mediated adsorption-desorption isotherm analysis using ASAP 2010, Micromeritics.

For determining light absorbance behavior of TiO$_2$ sample, Ultraviolet- Visible- Near IR (UV-Vis-NIR) spectroscopy was performed using Agilent Cary 5000i spectrophotometer.

4.4 Photocatalytic degradation of pharmaceutical pollutants

Photocatalytic degradation of three aqueous pharmaceutical pollutants (LFX, MNZ and KRL) was carried out in photocatalytic reactor with cooling jacket under constant stirring condition with help of magnetic stirring system. Under the UV light source (Phillips high pressure mercury lamp of 125W), 350 ml aqueous pollutant solution was taken with 50 mg/L initial pollutant concentration and 1 g/L TiO$_2$ catalyst for all three targeted pharmaceutical pollutants. For LFX, influence of catalyst loading (0.5-2 g/L) under constant pollutant concentration was studied for studying photoactivity behavior. Also, effect of initial LFX concentration (25-100 mg/L) on degradation performance at constant catalyst loading was checked. The schematic diagram of photocatalytic reactor is shown in Fig. 2.

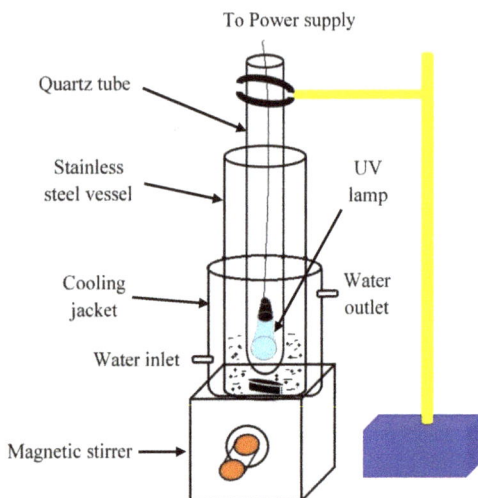

Figure 2. Schematic diagram of photocatalytic reactor.

During the photocatalytic reaction, the sample was taken out for find out the concentration of aqueous pollutant at a 15 min. time interval, which was centrifuged for separating TiO$_2$ nanomaterials. The absorption spectra profile of pharmaceutical aqueous solutions was analysed in Agilent Cary 5000i spectrophotometer. The rate of degradation was determined through change in λ_{max} of targeted pollutants. LFX, KRL and MNZ have exhibited maximum absorbance at 288 mm, 323 nm and 319 nm respectively. Calculation of degradation percentage is mentioned as in Eq. 11.

% Degradation = {(C$_0$-C)/C$_0$} × 100   (11)

Where, C$_0$ is initial concentration of pharmaceutical solution and C is concentration of pharmaceutical solution after UV light irradiation.

5. Result and discussions

5.1 TiO₂ photocatalyst characteristics

5.1.1 TGA of Ti(OH)₄

Thermal profile of Ti(OH)₄ sample is shown in Fig. 3, where weight losses was noticed at different stages. In first stage, physisorbed water and solvents were removed at temperature range from 25°C to around 250°C. Removal of surface attached surfactant and dihydroxylation of Ti(OH)₄ are could be attributed to weight loss in second stage (260°C to 350°C). Phase transformation of TiO₂ from amorphous to crystalline may be occurred after second stage [20,51].

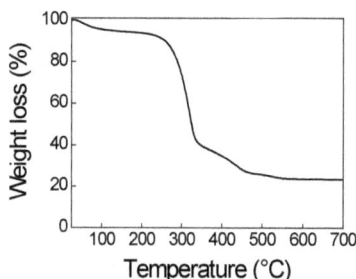

Figure 3. TGA profile of Ti(OH)₄ sample.

5.1.2 X-ray diffraction

The diffraction pattern of TiO₂ nanomaterials is shown in Fig. 4, where diffraction peaks at $2\theta = 25.3°, 37.9°, 48.1°, 54.1°, 55.1°$ and $62.7°$ are related to (1 0 1), (0 0 4), (2 0 0), (1 0 5), (2 1 1), and (2 0 4) orientations of TiO₂ anatase phase (ICDD No.01-089-4921) which is denoted as A. Rutile phase peak is also observed at $2\theta = 27.4°$ corresponded to (1 1 0) phase (ICDD No. 01-083-2242) which is symbolized as R [52].

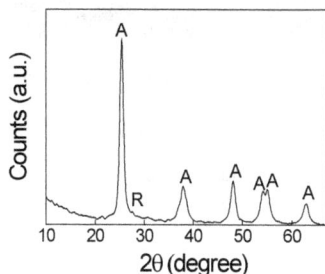

Figure 4. XRD pattern of TiO₂ sample.

Based on XRD analysis of TiO_2 nanomaterials, anatase and rutile phase are found with 95.15% and 4.85% respectively. As per the Scherrer's equation, average crystallite size of TiO_2 is estimated as 7.92 nm. Under the reverse microemulsion environment, Hong et al. [53] are utilized Brij 58 surfactant-cyclohexane based method for synthesizing TiO_2 nanomaterials, where only anatase phase for TiO_2 was observed which is calcined at 400°C and crystallite size was 8 nm.

5.1.3 Scanning electron microscopy

The SEM picture as shown in Fig. 5 is suggested that synthesized TiO_2 nanomaterials through reverse micelle mediated sol-gel method is spherical in shape and also PSD of nanomaterials are narrow.

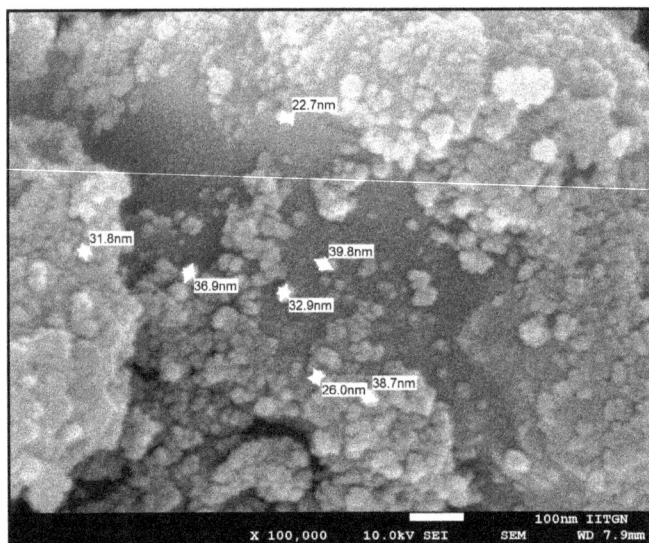

Figure 5. SEM image of TiO_2 sample.

5.1.4 Particle size using dynamic light scattering

Based on the DLS analysis, the number of average particle size of TiO_2 nanomaterial is found as 36.8 nm. It is also observed that distribution of particle size is narrow as shown in Fig. 6.

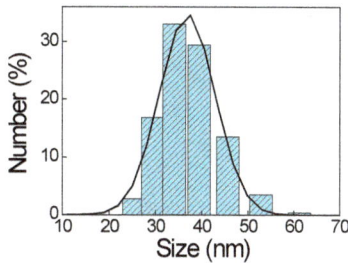

Figure 6. Number (%) vs. particle size of TiO₂ sample.

5.1.5 Surface area study

BET surface area and surface properties of TiO_2 nanomaterials is carried out using N_2 gas as an adsorbate under the bath temperature of -195°C. BET surface area of TiO_2 sample is found as 77.1 m²/g with pore volume of 0.1255 cm³/g and pore size of 65.13 Å. N_2 adsorption and desorption for TiO_2 sample is shown as in Fig. 7. Based on analysis, it is observed that from the different isotherm nature, TiO_2 nanomaterials have well matched with type IV isotherm behavior conferring to IUPAC classification. Moreover, type H2 hysteresis loop is identical to adsorption isotherm loop for TiO_2 nanomaterials. This adsorption behavior suggested that nanomaterials have mesoporous structure [54-56].

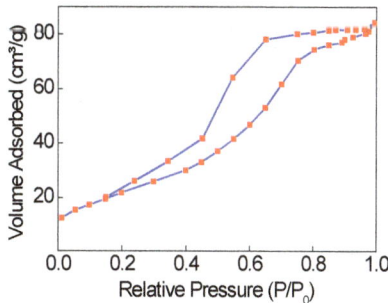

Figure 7. N₂ adsorption-desorption isotherm plot of TiO₂ sample.

5.1.6 UV-visible spectroscopy

UV-vis spectra of TiO_2 solid sample is carried out in wavelength range of 200-800 nm which is shown in Fig. 8a. The tauc plot of light energy $(\alpha h v)^2$ versus energy (hv) is shown in Fig. 8b, where it is estimated that bang gap energy of TiO_2 sample is around

3.19 eV. Based on reported study of TiO$_2$ nanomaterials synthesized through sol-gel reverse micelle method, it is found that band gap energy of TiO$_2$ nanomaterials is 3.13 eV [57].

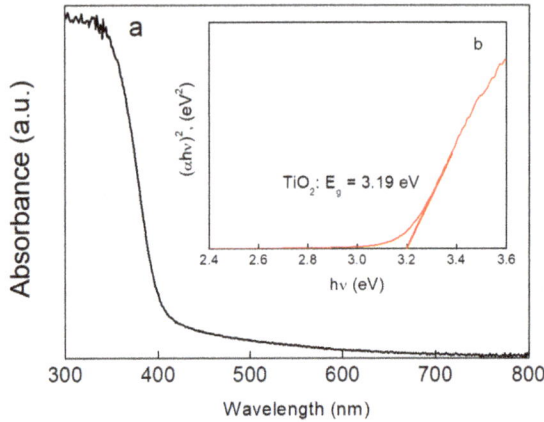

Figure 8. a) Absorbance spectra of TiO$_2$ sample, b) Tauc plot.

5.2 Influence of factors on photocatalytic activity of TiO$_2$ against LFX

5.2.1 Effect of catalyst dose

For levofloxacin (50 mg/L) degradation study under UV light source, effect of TiO$_2$ photocatalyst loading is checked. When amount of catalyst is increased from 0.5 g/L to 1 g/L, removal efficiency of LFX is increased from 97% to 99% as shown in Fig. 9a. Performance is not much enhanced for 2 g/L catalyst, which may be resulted due to increment in opacity and lesser number of active sites in photoreactor resulted from higher catalyst loading. Sharma et al. [58] is synthesized TiO$_2$/C-dots composite nanomaterials for levofloxacin degradation and checked the effect of catalyst loading on removal efficiency under solar light. It is found that optimum catalyst loading is 0.25 g/L and at higher photocatalyst amount of 0.5 g/L degradation efficiency is decreased. Similar kind of results of lowering in degradation efficiency with increment in catalyst amount is observed, where (BiOBr)$_x$(Bi$_7$O$_9$I$_3$)$_{1-x}$ nanostructure material is utilized for levofloxacin degradation [37].

The kinetics of photocatalytic degradation of LVF is studied by using Langmuir–Hinshelwood model and its rate equation typified as Eq. 12 [20,59].

Advances in Wastewater Treatment I
Materials Research Foundations **91** (2021) 87-110

Materials Research Forum LLC
https://doi.org/10.21741/9781644901144-3

$$-\ln (C/C_0) = kt \qquad (12)$$

Where C_0 is initial concentration of LFX aqueous solution at time $t = 0$ and C is concentration of LFX aqueous solution at different photocatalytic reaction time t. Slope of straight line is provided the rate constant k.

Plot of $-\ln(C/C_0)$ vs. time is illustrated as in Fig. 9b. The rate constant for different catalyst amount of 0.5, 1 and 2 g/L is 0.0648, 0.1035 and 0.1258 min^{-1} respectively.

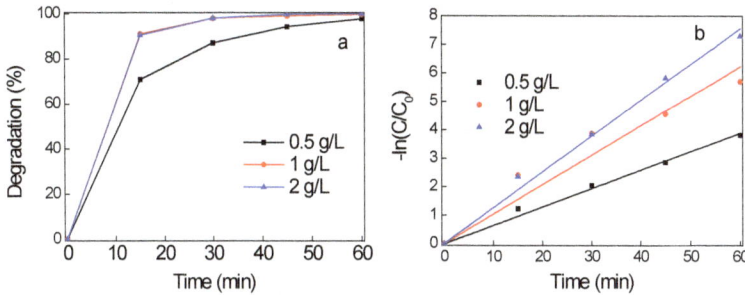

Figure 9. a) Effect of catalyst dose on degradation performance, b) Kinetics of LFX degradation (LFX concentration =50 mg/L, reaction time = 60 min).

5.2.2 Effect of initial concentration of LFX

Initial concentration of pollutant is directly influenced the degradation performance. In this study, for three different LFX concentration 25, 50 and 100 mg/L, photoactivity of synthesized TiO_2 nanomaterial is checked. When concentration of LFX is increased from 25 to 50 mg/L, photocatalytic performance is almost similar as shown in Fig. 10a. However, when 100 mg/L LFX aqueous solution is used, degradation efficiency is reached to 88%, which mostly due to over accumulation of LFX molecules which covered the active sites of TiO_2. Similar kind of declined trend in photoactivity was observed at higher initial pollutant concertation as per reported study, where TiO_2/Carbon-dots nanocomposite was utilized as photocatalyst [58].

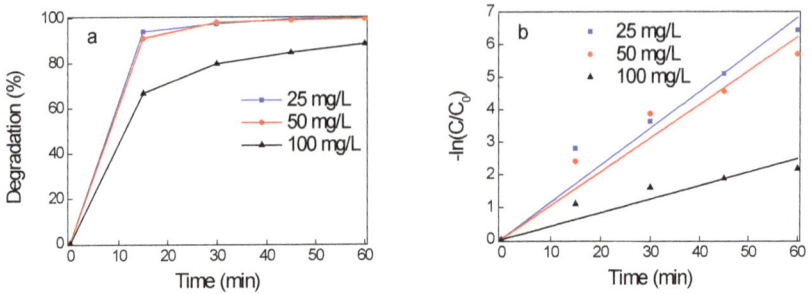

Figure 10. a) Effect of initial concentration of LFX on degradation performance, b)
Kinetics of LFX degradation (catalyst amount=1 g/L, reaction time = 60 min).

Degradation kinetics of LFX at different initial pollutant concentration is shown in Fig. 10b. The corresponding value of rate constant for 25, 50 and 100 mg/L LFX concentration is 0.1134, 0.1035 and 0.0414 min^{-1} respectively. It is observed that rate constant is decreases with increasing pollutant concentration.

5.3   Photocatalytic activity of TiO₂ and Degussa P25 nanomaterials against LFX, KRL and MNZ

Under UV light illumination, comparison of photocatalytic degradation performance of TiO₂ nanomaterial synthesized through reverse micelle method and commercial Degussa P25 (TiO₂) nanomaterials have carried out at optimum catalyst loading and initial pollutant concentration which was 1 g/L and 50 mg/L respectively. It was observed that for LFX and KRL pollutants, TiO₂ nanomaterial synthesized through reverse micelle method has shown equivalent degradation performance compared to Degussa P25 nanomaterial as shown in Fig. 11a. While, Degussa P25 nanomaterial has exhibited slightly higher degradation performance compared to synthesized TiO₂ nanomaterial for MNZ pollutant. Plot of degradation performance of synthesized and commercial TiO₂ nanomaterial in terms of C/C₀ versus time for all targeted pharmaceutical pollutants is given in Fig. 11b.

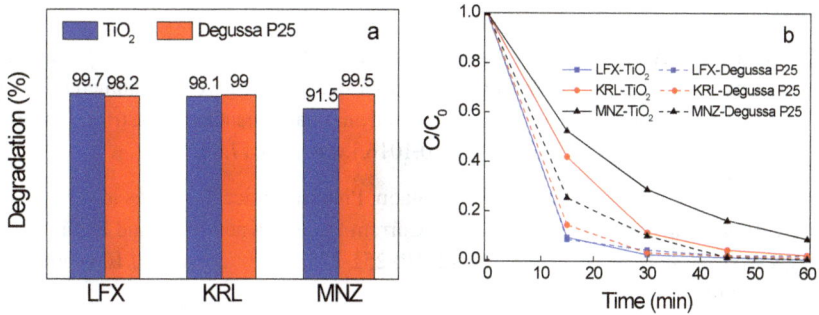

Figure 11. Plot of a) Degradation performance, b) C/C₀ vs. time for TiO₂ and Degussa P25 against LFX, KRL and MNZ (catalysts amount = 1 g/L, concentration of pollutants = 50 mg/L).

Conclusion

With utilization of reverse micelle based modified sol-gel method, TiO_2 has key features like narrow particle size distribution with size range of 20-40 nm. Also, synthesized TiO_2 nanomaterial has higher surface area and mesoporous characteristics. TiO_2 nanomaterial has shown significant performance against LFX, KRL and MNZ aqueous pharmaceutical pollutants under UV light irradiation. For LFX and KRL, synthesized TiO_2 nanomaterial has shown equivalent degradation performance to commercial Degussa P25 nanomaterial. Optimization study of process parameters for photocatalytic degradation of LFX has checked, where optimum catalyst loading and initial pollutant concentration was found to be as 1 g/L and 50 mg/L respectively. This experimental study suggested that TiO_2 nanomaterials synthesized through reverse micelle-based method has performed efficiently against pharmaceutical pollutants under UV light source.

Acknowledgements

Authors deeply acknowledge the Gujarat Pollution Control Board (GPCB), Gandhinagar, Gujarat (India) for providing Ph.D. fellowship (Grant ID: Ph.D./R&D/05/444437). The authors are grateful to Dharmsinh Desai University, Nadiad for providing essential research facilities and thankful to Dr. B. N. Suhagia, Dean, Faculty of Pharmacy, Dharmsinh Desai University for his valuable guidance in selection of pharmaceutical pollutants for photodegradation study.

References

[1]   A. Bielen, A. Šimatović, J. Kosić-Vukšić, I. Senta, M. Ahel, S. Babić, T. Jurina, J.J. González Plaza, M. Milaković, N. Udiković-Kolić, Negative environmental impacts of antibiotic-contaminated effluents from pharmaceutical industries, Water Res. 126 (2017) 79-87. https://doi.org/10.1016/j.watres.2017.09.019

[2]   D. Fatta-Kassinos, S. Meric, A. Nikolaou, Pharmaceutical Residues in Environmental Waters and Wastewater: Current State of Knowledge and Future Research, Anal. Bioanal. Chem. 399 (2010) 251-275. https://doi.org/10.1007/s00216-010-4300-9

[3]   Q. Bu, X. Shi, G. Yu, J. Huang, B. Wang, Assessing the persistence of pharmaceuticals in the aquatic environment: Challenges and needs, Emerg. Contam. 2 (2016) 145-147. https://doi.org/10.1016/j.emcon.2016.05.003

[4]   A. Jelić, M. Gros, M. Petrović, A. Ginebreda, D. Barceló, Occurrence and Elimination of Pharmaceuticals During Conventional Wastewater Treatment BT - Emerging and Priority Pollutants in Rivers: Bringing Science into River Management Plans, in: H. Guasch, A. Ginebreda, A. Geiszinger (Eds.), Springer Berlin Heidelberg, Berlin, Heidelberg, 2012: pp. 1-23. https://doi.org/10.1007/978-3-642-25722-3_1

[5]   P. Bottoni, S. Caroli, A.B. Caracciolo, Pharmaceuticals as priority water contaminants, Toxicol. Environ. Chem. 92 (2010) 549-565. https://doi.org/10.1080/02772241003614320

[6]   H. Zhou, D.W. Smith, Advanced technologies in water and wastewater treatment, J. Environ. Eng. Sci. 1 (2002) 247-264. https://doi.org/10.1139/s02-020

[7]   L.V. Bora, R.K. Mewada, Visible/solar light active photocatalysts for organic effluent treatment: Fundamentals, mechanisms and parametric review, Renew. Sustain. Energy Rev. 76 (2017) 1393-1421. https://doi.org/10.1016/j.rser.2017.01.130

[8]   E. GilPavas, I. Dobrosz-Gómez, M.Á. Gómez-García, Coagulation-flocculation sequential with Fenton or Photo-Fenton processes as an alternative for the industrial textile wastewater treatment, J. Environ. Manage. 191 (2017) 189-197. https://doi.org/10.1016/j.jenvman.2017.01.015

[9]   S. Loaiza-Ambuludi, M. Panizza, N. Oturan, M.A. Oturan, Removal of the anti-inflammatory drug ibuprofen from water using homogeneous photocatalysis, Catal. Today. 224 (2014) 29-33. https://doi.org/10.1016/j.cattod.2013.12.018

[10]   L. Fu, C. Wu, Y. Zhou, J. Zuo, G. Song, Y. Tan, Ozonation reactivity characteristics of dissolved organic matter in secondary petrochemical wastewater by single ozone, ozone/H$_2$O$_2$, and ozone/catalyst, Chemosphere. 233 (2019) 34-43. https://doi.org/10.1016/j.chemosphere.2019.05.207

[11]   J.C. Cardoso, G.G. Bessegato, M.V. Boldrin Zanoni, Efficiency comparison of ozonation, photolysis, photocatalysis and photoelectrocatalysis methods in real textile wastewater decolorization, Water Res. 98 (2016) 39-46. https://doi.org/10.1016/j.watres.2016.04.004

[12]   F. Martínez, M.J. López-Muñoz, J. Aguado, J.A. Melero, J. Arsuaga, A. Sotto, R. Molina, Y. Segura, M.I. Pariente, A. Revilla, L. Cerro, G. Carenas, Coupling membrane separation and photocatalytic oxidation processes for the degradation of pharmaceutical pollutants, Water Res. 47 (2013) 5647-5658. https://doi.org/10.1016/j.watres.2013.06.045

[13]   K. Natarajan, T. Natarajan, H. Bajaj, R. Tayade, High photocatalytic activity of rutile phase dominant TiO$_2$ synthesized by thermal treatment under narrow spectrum ultraviolet light emitting diodes, Mater. Res. Express. 6 (2018). https://doi.org/10.1088/2053-1591/aae861

[14] R.J. Tayade, W.K. Jo, Enhanced Photocatalytic Activity of TiO$_2$ Supported on Different Carbon Allotropes for Degradation of Pharmaceutical Organic Compounds, in: R.J. Tayade, V. Gandhi (Eds.), Photocatalytic Nanomaterials for Environmental Applications, Materials Research Forum LLC, Millersville, PA, 2018, pp. 140-159. http://dx.doi.org/10.21741/9781945291593-4

[15]   A. Eslami, M. Amini, A. Yazdanbakhsh, A. Mohseni-Bandpei, A. Safari, A. Asadi, N,S co-doped TiO$_2$ nanoparticles and nanosheets in simulated solar light for photocatalytic degradation of non-steroidal anti-inflammatory drugs in water: A comparative study, J. Chem. Technol. Biotechnol. 91 (2015) n/a-n/a. https://doi.org/10.1002/jctb.4877

[16]   S.S. Boxi, S. Paria, Visible light induced enhanced photocatalytic degradation of organic pollutants in aqueous media using Ag doped hollow TiO$_2$ nanospheres, RSC Adv. 5 (2015) 37657-37668. https://doi.org/10.1039/C5RA03421C

[17]   R. Fagan, D.E. McCormack, D.D. Dionysiou, S.C. Pillai, A review of solar and visible light active TiO$_2$ photocatalysis for treating bacteria, cyanotoxins and contaminants of emerging concern, Mater. Sci. Semicond. Process. 42 (2016) 2-14. https://doi.org/10.1016/j.mssp.2015.07.052

[18]   C. Byrne, G. Subramanian, S.C. Pillai, Recent advances in photocatalysis for environmental applications, J. Environ. Chem. Eng. 6 (2018) 3531-3555. https://doi.org/10.1016/j.jece.2017.07.080

[19]   W.H.M. Abdelraheem, M.K. Patil, M.N. Nadagouda, D.D. Dionysiou, Hydrothermal synthesis of photoactive nitrogen- and boron- codoped TiO2 nanoparticles for the treatment of bisphenol A in wastewater: Synthesis, photocatalytic activity, degradation byproducts and reaction pathways, Appl. Catal. B Environ. 241 (2019) 598-611. https://doi.org/10.1016/j.apcatb.2018.09.039

[20]   V.G. Gandhi, M. Mishra, S. Meka, A. Kumar, P. Joshi, D. Shah, Comparative study on nano-crystalline titanium dioxide catalyzed photocatalytic degradation of aromatic carboxylic acids in aqueous medium, J. Ind. Eng. Chem. 17 (2011) 331-339. https://doi.org/10.1016/j.jiec.2011.02.035

[21]   N. Venkatachalam, M. Palanichamy, V. Murugesan, Sol–gel preparation and characterization of nanosize TiO_2: Its photocatalytic performance, Mater. Chem. Phys. 104 (2007) 454-459. https://doi.org/10.1016/j.matchemphys.2007.04.003

[22]   R.S. Dubey, K.V. Krishnamurthy, S. Singh, Experimental studies of TiO_2 nanoparticles synthesized by sol-gel and solvothermal routes for DSSCs application, Results Phys. 14 (2019) 102390. https://doi.org/10.1016/j.rinp.2019.102390

[23]   J.-K. Oh, J.-K. Lee, S.J. Kim, K.-W. Park, Synthesis of phase- and shape-controlled TiO_2 nanoparticles via hydrothermal process, J. Ind. Eng. Chem. 15 (2009) 270-274. https://doi.org/10.1016/j.jiec.2008.10.001

[24]   M.S. Lee, S.S. Park, G.-D. Lee, C.-S. Ju, S.-S. Hong, Synthesis of TiO_2 particles by reverse microemulsion method using nonionic surfactants with different hydrophilic and hydrophobic group and their photocatalytic activity, Catal. Today. 101 (2005) 283-290. https://doi.org/10.1016/j.cattod.2005.03.018

[25]   C. Dhand, N. Dwivedi, X.J. Loh, A.N. Jie Ying, N.K. Verma, R.W. Beuerman, R. Lakshminarayanan, S. Ramakrishna, Methods and strategies for the synthesis of diverse nanoparticles and their applications: a comprehensive overview, RSC Adv. 5 (2015) 105003–105037. https://doi.org/10.1039/C5RA19388E

[26]   P. Arabkhani, S. Abedini-Khorrami, Synthesis of $SrAl_2O_4$ nanoparticles by reverse micelle method: An investigation of the effect of aging time, hydrophilic chain length and calcination temperature, Nano-Structures and Nano-Objects. 17 (2019) 84-91. https://doi.org/10.1016/j.nanoso.2018.12.003

[27] B. Richard, J.-L. Lemyre, A.M. Ritcey, Nanoparticle Size Control in Microemulsion Synthesis, Langmuir. 33 (2017) 4748-4757. https://doi.org/10.1021/acs.langmuir.7b00773

[28] L. Hu, P.M. Flanders, P.L. Miller, T.J. Strathmann, Oxidation of sulfamethoxazole and related antimicrobial agents by TiO_2 photocatalysis, Water Res. 41 (2007) 2612-2626. https://doi.org/10.1016/j.watres.2007.02.026

[29] A.L. Giraldo, G.A. Peñuela, R.A. Torres-Palma, N.J. Pino, R.A. Palominos, H.D. Mansilla, Degradation of the antibiotic oxolinic acid by photocatalysis with TiO_2 in suspension, Water Res. 44 (2010) 5158-5167. https://doi.org/10.1016/j.watres.2010.05.011

[30] Y. Lin, C. Ferronato, N. Deng, J.-M. Chovelon, Study of benzylparaben photocatalytic degradation by TiO_2, Appl. Catal. B Environ. 104 (2011) 353-360. https://doi.org/10.1016/j.apcatb.2011.03.006

[31] N. Laoufi, S. Hout, T. Djilali, A. Ounnar, A. Djouadi, N. Chekir, F. Bentahar, Removal of a Persistent Pharmaceutical Micropollutant by UV/TiO_2 Process Using an Immobilized Titanium Dioxide Catalyst: Parametric Study, Chem. Eng. Trans. 32 (2013) 1951-1956. https://doi.org/10.3303/CET1332326

[32] J.L. WANG, L.E.J.I.N. XU, Advanced Oxidation Processes for Wastewater Treatment: Formation of Hydroxyl Radical and Application, Crit. Rev. Environ. Sci. Technol. 42 (2012) 251-325. https://doi.org/10.1080/10643389.2010.507698

[33] K.S. Varma, R.J. Tayade, K.J. Shah, P.A. Joshi, A.D. Shukla, V.G. Gandhi, Photocatalytic degradation of pharmaceutical and pesticide compounds (PPCs) using doped TiO_2 nanomaterials: A review, Water-Energy Nexus. 3 (2020) 46-61. https://doi.org/10.1016/j.wen.2020.03.008

[34] O. Dalrymple, D. Yeh, M. Trotz, Removing Pharmaceuticals and Endocrine-Disrupting Compounds from Wastewater by Photocatalysis, J. Chem. Technol. Biotechnol. 82 (2007) 121-134. https://doi.org/10.1002/jctb.1657

[35] K.M. Reza, A.S.W. Kurny, F. Gulshan, Parameters affecting the photocatalytic degradation of dyes using TiO_2: a review, Appl. Water Sci. 7 (2017) 1569-1578. https://doi.org/10.1007/s13201-015-0367-y

[36] V. Gandhi, M. Mishra, P.A. Joshi, Titanium dioxide catalyzed photocatalytic degradation of carboxylic acids from waste water: A review, Mater. Sci. Forum. 712 (2012) 175-189. https://doi.org/10.4028/www.scientific.net/MSF.712.175

[37]   S.G. Fard, M. Haghighi, M. Shabani, Facile one-pot ultrasound-assisted solvothermal fabrication of ball-flowerlike nanostructured $(BiOBr)_x(Bi_7O_9I_3)_{1-x}$ solid-solution for high active photodegradation of antibiotic levofloxacin under sun-light, Appl. Catal. B Environ. 248 (2019) 320–331. https://doi.org/10.1016/j.apcatb.2019.02.021

[38]   R.S. Dariani, A. Esmaeili, A. Mortezaali, S. Dehghanpour, Photocatalytic reaction and degradation of methylene blue on TiO_2 nano-sized particles, Optik (Stuttg). 127 (2016) 7143-7154. https://doi.org/10.1016/j.ijleo.2016.04.026

[39]   G.H. Safari, M. Hoseini, M. Seyedsalehi, H. Kamani, J. Jaafari, A.H. Mahvi, Photocatalytic degradation of tetracycline using nanosized titanium dioxide in aqueous solution, Int. J. Environ. Sci. Technol. 12 (2015) 603-616. https://doi.org/10.1007/s13762-014-0706-9

[40]   A. Hassani, A. Khataee, S. Karaca, Photocatalytic degradation of ciprofloxacin by synthesized TiO_2 nanoparticles on montmorillonite: Effect of operation parameters and artificial neural network modeling, J. Mol. Catal. A Chem. 409 (2015) 149-161. https://doi.org/10.1016/j.molcata.2015.08.020

[41]   H. Yang, G. Li, T. An, Y. Gao, J. Fu, Photocatalytic degradation kinetics and mechanism of environmental pharmaceuticals in aqueous suspension of TiO_2: A case of sulfa drugs, Catal. Today. 153 (2010) 200-207. https://doi.org/10.1016/j.cattod.2010.02.068

[42]   M. Malakootian, A. Nasiri, M. Gharaghani, Photocatalytic degradation of ciprofloxacin antibiotic by TiO_2 nanoparticles immobilized on a glass plate, Chem. Eng. Commun. (2019) 1-17. https://doi.org/10.1080/00986445.2019.1573168

[43]   A. Nickheslat, M.M. Amin, H. Izanloo, A. Fatehizadeh, S.M. Mousavi, Phenol Photocatalytic Degradation by Advanced Oxidation Process under Ultraviolet Radiation Using Titanium Dioxide, J. Environ. Public Health. 2013 (2013) 815310. https://doi.org/10.1155/2013/815310

[44]   J.R. Alvarez-Corena, J.A. Bergendahl, F.L. Hart, Advanced oxidation of five contaminants in water by UV/TiO_2: Reaction kinetics and byproducts identification, J. Environ. Manage. 181 (2016) 544-551. https://doi.org/10.1016/j.jenvman.2016.07.015

[45]   J. Vyas, M. Mishra, V. Gandhi, Photocatalytic Degradation of Alizarin Cyanine Green G, Reactive Red 195 and Reactive Black 5 Using UV/TiO_2 Process, Mater. Sci. Forum. 764 (2013) 284-292. https://doi.org/10.4028/www.scientific.net/MSF.764.284

[46] R. Patel, T. Bhingradiya, A. Deshmukh, V. Gandhi, Response Surface Methodology for Optimization and Modeling of Photo-Degradation of Alizarin Cyanine Green and Acid Orange 7 Dyes Using UV/TiO$_2$ Process, Mater. Sci. Forum. 855 (2016) 94-104. https://doi.org/10.4028/www.scientific.net/MSF.855.94

[47] A. Carabin, P. Drogui, D. Robert, Photo-degradation of carbamazepine using TiO$_2$ suspended photocatalysts, J. Taiwan Inst. Chem. Eng. 54 (2015) 109-117. https://doi.org/10.1016/j.jtice.2015.03.006

[48] M. Faramarzpour, M. Vossoughi, M. Borghei, Photocatalytic degradation of furfural by titania nanoparticles in a floating-bed photoreactor, Chem. Eng. J. 146 (2009) 79-85. https://doi.org/10.1016/j.cej.2008.05.033

[49] S. Mathew, P. Ganguly, S. Rhatigan, V. Kumaravel, C. Byrne, S.J. Hinder, J. Bartlett, M. Nolan, S.C. Pillai, Cu-Doped TiO$_2$: Visible Light Assisted Photocatalytic Antimicrobial Activity, Appl. Sci. 8 (2018). https://doi.org/10.3390/app8112067

[50] R. Bashiri, N.M. Mohamed, C.F. Kait, S. Sufian, Hydrogen production from water photosplitting using Cu/TiO$_2$ nanoparticles: Effect of hydrolysis rate and reaction medium, Int. J. Hydrogen Energy. 40 (2015) 6021-6037. https://doi.org/10.1016/j.ijhydene.2015.03.019

[51] H. Benito, T. Sánchez, R. Alamilla, J. Hernandez, G. Robles, F. Paraguay-Delgado, Synthesis and physicochemical characterization of titanium oxide and sulfated titanium oxide obtained by thermal hydrolysis of titanium tetrachloride, Brazilian J. Chem. Eng. 31 (2014) 737-745. https://doi.org/10.1590/0104-6632.20140313s00002506

[52] P. Singla, O.P. Pandey, K. Singh, Study of photocatalytic degradation of environmentally harmful phthalate esters using Ni-doped TiO$_2$ nanoparticles, Int. J. Environ. Sci. Technol. 13 (2016) 849-856. https://doi.org/10.1007/s13762-015-0909-8

[53] S.-S. Hong, M.S. Lee, G.-D. Lee, Photocatalytic decomposition of p-nitrophenol over titanium dioxide prepared by reverse microemulsion method using nonionic surfactants with different hydrophilic groups, React. Kinet. Catal. Lett. 80 (2003) 145-151. https://doi.org/10.1023/A:1026096628817

[54] M. Naderi, Surface Area: Brunauer–Emmett–Teller (BET), in: S. Tarleton (Eds.), Progress in Filtration and Separation, Academic Press, Oxford, 2015, pp. 585-608. https://doi.org/10.1016/B978-0-12-384746-1.00014-8

[55] M. Thommes, Physical Adsorption Characterization of Nanoporous Materials, Chemie Ing. Tech. 82 (2010) 1059–1073. https://doi.org/10.1002/cite.201000064

[56] S.M. Reda, M. Khairy, M.A. Mousa, Photocatalytic activity of nitrogen and copper doped TiO_2 nanoparticles prepared by microwave-assisted sol-gel process, Arab. J. Chem. 13 (2020) 86-95. https://doi.org/10.1016/j.arabjc.2017.02.002

[57] K. Elghniji, M. Ksibi, E. Elaloui, Sol–gel reverse micelle preparation and characterization of N-doped TiO_2: Efficient photocatalytic degradation of methylene blue in water under visible light, J. Ind. Eng. Chem. 18 (2012) 178-182. https://doi.org/10.1016/j.jiec.2011.11.011

[58] S. Sharma, A. Umar, S. Mehta, A. Ibhadon, S. Kansal, Solar light driven photocatalytic degradation of levofloxacin using TiO_2/Carbon-dots nanocomposite, New J. Chem. 42 (2018). https://doi.org/10.1039/C7NJ05118B

[59] A. Kaur, A. Umar, S.K. Kansal, Sunlight-driven photocatalytic degradation of non-steroidal anti-inflammatory drug based on TiO_2 quantum dots, J. Colloid Interface Sci. 459 (2015) 257-263. https://doi.org/10.1016/j.jcis.2015.08.010

Materials Research Forum LLC
https://doi.org/10.21741/9781644901144-4

Chapter 4

Treatment and Analysis of Arsenic Contaminated Water

Arup K. Ghosh[1]*

[1]Department of Chemical Engineering (Chemistry), Dharmsinh Desai University, College Road, Nadiad-387001, Gujarat, India

arupchem87@gmail.com

Abstract

The presence of arsenic in as many as 245 minerals makes it an indispensable waste in the metal refining industry. Hydraulic fracturing, underground drilling, pesticides, herbicides, electronic industries are also linked to arsenic contamination. Natural processes such as volcanic emissions, hydrothermal ores, and river flow through arsenic rich sediments also contribute to arsenic contaminated water. The consumption of arsenic contaminated water leads to various types of cancer such as dermatological, respiratory, gastrointestinal, cardiovascular, hepatic, neurological, renal, and mutagenesis. Thus, remediation and testing of arsenic contaminated water becomes ubiquitous. Arsenic removal methods include precipitation, filtration, membrane technology and bioremediation. Quantitative arsenic analysis includes several colorimetric, luminescence, spectroscopic, atomic absorption, mass spectrometric and biosensor-based techniques. In this chapter, we present an overview of the various sources linked with arsenic contaminated water followed by a discussion on the available treatment and monitoring technologies for waterborne arsenic.

Keywords

Arsenic, Water Treatment, Arsenic Analysis, Groundwater, Sorption, Membrane Filtration, Colorimetry, Gas Phase Chemiluminescence

Contents

Materials Research Forum LLC
https://doi.org/10.21741/9781644901144-4

1.   Introduction

Arsenic (As) with an atomic number of 33, belongs to group 15 of the modern periodic table. It has an electronic configuration of $[Ar]3d^{10}\ 4s^2\ 4p^3$ and shows several oxidation states which include -3, 0, +3, and +5. As many as 29 isotopes of arsenic have been discovered, among which ^{75}As is the only stable and naturally occurring isotope [1]. In addition to the presence elemental arsenic, arsenic itself also combines with many other elements to form as many as 245 minerals [2]. The major arsenic containing minerals include arsenides, arsenosulfides, arsenites and arsenates. Less known forms of arsenic compounds include sulfide, sulfate, carbonate and even organoarsenicals.

Many of these arsenic compounds have been used by the human civilizations [2,3]. Arsenic compounds such as realgar (As_4S_4), orpiment (As_2S_3),and arsenolite (As_4O_6) have been have been utilized for making pigments, medicines, alloys, pesticides, herbicides, glassware, embalming fluids, and also for removing unwanted hair in the leather manufacturing industry [3]. Furthermore, the well-known toxic properties of arsenic compounds have been widely used in traditional medicines, chemical warfare agents, to commit murder and even suicide [2,3]. Moreover, arsenic was utilized in livestock dips, feed supplements, and wood preservatives [3]. Recently, with the growing demand of various electronic devices such as light emitting diodes (LEDs), lasers, photodiodes, mobile phones and navigation systems, the use of gallium arsenide (GaAs) and indium arsenide (InAs) in the semiconductor industry has amplified over the past few decades [3-7]. However, with the increasing concern on the adverse health effects of arsenic exposure found among the workers, health agencies have emphasized upon the reduction in the industrial use of arsenic [7-8].

Arsenic has been associated with wide range of health problems. Inorganic As(III) and As(V) compounds, such as arsenic trioxide (As_2O_3), arsenites (AsO_3^-) and arsenates (AsO_4^-) are known to cause several kinds of cancer which include skin, kidney, urinary bladder, lung, and prostate cancer [3-6]. Other serious health issues related to arsenic include diabetes, keratosis, hyperpigmentation, liver failure, cardiovascular, hepatic, renal, gastrointestinal, neurological, reproductive, chromosomal and DNA damage [2-7]. Moreover, these diseases have led to many social stigmatization issues in developing countries like India and Bangladesh, where most of these diseases have been related to

Materials Research Forum LLC
https://doi.org/10.21741/9781644901144-4

drinking arsenic contaminated water [7-8]. Several other cases of arsenic poisoning due to consumption of inorganic arsenic on a large scale are also well documented [7-8]. For instance, the case of arsenic poisoning of the Morinaga Milk in several cities of Japan in the year 1955 affected more than 12000 infants, including 130 deaths [8]. Organoarsenicals such as dimethyarsinic acid (DMA) and monomethylarsenoic acid (MMA) are known to be less harmful to humans as they are assimilated from the body through urine [9]. In this context, it is necessary to highlight the fact that the main path source of arsenic related health hazards are caused due to the consumption of arsenic contaminated food and groundwater. As per the recommendation of World Health Organization (WHO) guideline value, the maximum permissible level of arsenic in drinking water is 10 µg/L (1 µg/L = 1 parts per billion = 1 ppb) [10]. However, in many countries around the world such as Mexico, Mongolia, India, Bangladesh, Nepal, Taiwan, China and Vietnam, and even the United States, there are places where the arsenic level in groundwater is more than 50 µg/L [7]. It is estimated that all over the world, more than 42 million people are exposed to drinking water where arsenic levels are in excess of 50 µg/L, with more than 100 million people believed to be exposed to As concentrations above 10 µg/L [7].

Considering the abovementioned health hazards related to arsenic exposure, measures for removal and testing of waterborne arsenic becomes ubiquitous. Major arsenic removal technologies are centred around the principles of oxidation, precipitation/co-precipitation, sorption, ion-exchange and membrane process [11-137]. Other technologies, being used for treatment of arsenic contaminated water and soil include use of magnetic and biological treatment [138-166]. The quantitative measurement of arsenic in water samples is done using colorimetric, spectroscopic, atomic absorption, electrochemical, neutron activation, mass spectrometric methods and biosensors [167-267]. Recently, some cheaper test kits and methodologies have been also developed for field testing [167-171,210-212]. All these methods have their own benefits and restrictions. Several books and review articles have made attempts to elaborately address the treatment methods for arsenic present in water [3,11,33,40,50,82,84]. Similarly, arsenic analysis techniques have also been reviewed separately many times [200,217,227-228,257-258,268-270]. In this chapter, a concise effort has been made to bring together some of the major available arsenic treatment technologies and analytical methods with an aim to create awareness and cease the flow of arsenic in the immediate human environment.

2. Sources of arsenic

Arsenic is the 20th most abundant element in the earth's crust and hence, its presence in various rocks, metal ores, concentrates, coal, oil, chemical and industrial products, and its flow through the ecosystem is very much probable [2,3]. Arsenic release in the environment can be divided primarily in two main categories. Firstly, natural processes which include weathering of rocks and sediments, hydrothermal fluids, ore deposits, volcanic eruptions, geothermal activities, forest fire, etc [2,3]. Secondly, anthropogenic sources consisting of human and industrial activities such as urban and industrial waste, mining, fracking, smelting, ore processing, pesticides, fertilizers, chemical industries, thermal power plants using coal and wood preservation industries [2,3]. Below, a brief discussion on the abovementioned categories is presented to provide a better understanding of the merits and demerits associated with the available removal and analytical techniques available for arsenic.

2.1   Natural processes

Natural sources of arsenic include weathering and leaching from arsenic rich deposits, volcanoes, hydrothermal fluids, sea-salt aerosols, and forest fires. It is estimated that the mean annual global atmospheric emission of arsenic from such natural sources is more than 31000 metric tons [271]. Volcanoes are one of the major contributors of arsenic pollution in the atmosphere [3]. Orpiment (As_2S_3) and realgar (As_4S_4) are the common minerals associated with the volcanic emissions [2,3]. Hydrothermal fluids are naturally occurring hot waters which generally occur in the form of hot springs and natural geysers on the surface. These hydrothermal fluids often related with volcanic, metamorphic, and sedimentary rocks are known to bring arsenic to the earth's surface. The fact that solubility of arsenic increases in hot water, enables hydrothermal fluids to wash away arsenic and its compounds from sediments and rocks [3]. The average value of arsenic in sea water is approximately 2 parts per billion (ppb), which makes it the tenth most abundant element in sea water [3]. In sea water, arsenic exists mostly in the form of inorganic arsenic As(V) and As(III), whereas the organoarsenicals, MMA and DMA contribute less than 10% of the total arsenic present in the sea water [3]. Sea-salt aerosols are also known to increase the atmospheric arsenic [3]. Under reducing conditions, soils, fungi and other microorganisms are known to covert inorganic As(III) in to arsine (AsH_3) gas and methylated arsenic compounds (such as MMA and DMA) which are quite volatile [3]. Plants and trees absorb arsenic from the soil through water and when these plants are burnt, or during forest fires, significant amount of arsenic also released in the atmosphere [3].

Earth's crust contains many arsenic containing minerals such as arsenopyrite (FeAsS), orpiment (As_2S_3), and realgar (As_4S_4) [2-3]. Geological processes, such as weathering of rocks and minerals followed by subsequent leaching and runoff are significantly important, as these are associated with soil and groundwater contamination. However, there are many factors that affect the arsenic concentration and transport in groundwater, which include redox potential, adsorption/desorption rate, precipitation/dissolution, speciation, acidity/basicity, nature of competing ions, biological transformation, etc. which may vary from one aquifer to another depending upon the geo-chemistry and geological environment of the surroundings [3]. Hence, for sufficient knowledge of geological contribution to the arsenic contamination, a thorough analysis of the various hydrogeological and geo-environmental surroundings of the aquifers is very much necessary while planning for long term remedies.

2.2   Anthropogenic sources

Overall contributions to environmental arsenic from anthropogenic emissions are estimated at more than 23000 metric tonnes per year [271]. Anthropogenic sources of arsenic mainly consist of industrial activities like mining of arsenic rich ores, mineral processing, semiconductor industry and coal combustion. Moreover, agricultural use of herbicide and pesticides, wood preservation, cement, glass and steel industry are also known to release significant amounts of arsenic in the ecosystem [3].

One of the most important anthropogenic sources of arsenic is mine tailing. Mine tailings, following the oxidation of arsenic rich ores such as, arsenopyrite (FeAsS) contribute to arsenic in soil and water. Although, $Fe(OH)_3$ apparently provide a useful sink as arsenic gets adsorbed on the $Fe(OH)_3$ particles, however, changes in pH and redox conditions cause $Fe(OH)_3$ to dissolve and subsequently discharge arsenic in the environment [3]. Arsenic is also a common industrial waste produced during the smelting process of copper and several other non-iron metals such as silver, cobalt, nickel, lead, gold, zinc, manganese and tin [3]. In recent times, industries have been involved in production of lasers, photodiodes, light emitting diodes, cell phones which involve arsenic based semiconductor substances such as GaAs and InAs [4-5]. Being precious, gallium (Ga) as well as indium (In) are generally recycled and during this recycling process sufficient amounts of arsenic is converted into volatile arsine gas. This arsine gas poses a health hazard for the workers in the semiconductor industry as well as, the risk of any leakage may put the entire ecosystem in danger [4-5]. Combustion of arsenic rich coal in the industries is another source of volatile arsenic compounds and fine-grained arsenic containing aerosols in the atmosphere [3]. Moreover, coal mining has been found to be linked with high levels of arsenic in the stream sediments of Alabama, USA [3].

During the last couple of centuries, inorganic arsenic in the form of arsenic trioxide and arsenates of various metals have been widely used in pigments, pesticides, insecticides, herbicides and fungicides. Insecticides and pesticides containing arsenic have become a diffuse source of arsenic contaminated water mainly due to agricultural practices [3]. In many places, organoarsenicals such as, MMA and DMA were used as herbicides for maintenance of lawns and golf courses [3]. Chromated copper arsenate ($CrCuAs_2O_9$), commonly known as CCA and some other arsenic based chemicals, due to their biocidal properties, have been widely used as wood preservatives and have resulted in arsenic contaminated water and soil on a large scale [3]. In the steel industry, high concentrations of arsenic, along with cadmium (Cd), copper (Cu), iron (Fe), lead (Pb), nickel (Ni) and zinc (Zn) are quite common in the acid mine drainage discharges. Moreover, arsenic in its elemental form is primarily used in the alloys of Pb, Cu, antimony (Sb), tin (Sn), aluminium (Al), and gallium (Ga) [3].

3.   Treatment methodologies of arsenic contaminated water

The occurrence of arsenic in a large number of natural substances, industrial wastes and already contaminated areas, makes it indispensable to think about the possible remediation of arsenic contaminated sources. As already mentioned, arsenic contamination can take place in surface water, groundwater, soil and air, it becomes omnipotent that all the potential sources of arsenic must be treated to stop/reduce any kind of health hazard. Atmospheric concentration of arsenic in the form of particulate matter and volatile gases is relatively low. Moreover, atmospheric arsenic eventually settles down over a period of time to pollute soil and water [3]. Even arsenic from contaminated soils and oresis often leached out by run-off water, thus polluting surface water and seeps into groundwater as well. Thus, any form of arsenic release eventually leads to water contamination which exposes millions of people worldwide to various diseases [3-6]. Considering the millions of vulnerable people, the need for efficient yet inexpensive arsenic removal technologies from water is imminent.

Due to inaccessibility, treatment of groundwater without bringing it to the surface, is often more expensive compared to that of surface water. Surface waters, where the pH is primarily in the range 4-9, contains mostly As(V) [84]. In groundwater, where the conditions are less oxidizing contain significant amount of relatively more toxic As(III) along with As(V) [3]. Treatment of arsenic contaminated water is achieved through a variety of techniques such as sorption, ion exchange, precipitation, filtration, selective membranes, biological methods, and natural remediation [11-166]. However, efficiency of a particular method is greatly influenced by several factors namely, form of arsenic species, concentration, pH, presence or absence of other interfering ions and organic contaminants [3]. These factors determine the amount of water required for treatment, expenditures, and the fact that where the treated water is going to be used. Sometimes two or more of the abovementioned methods are used in combination to achieve the desired results. Selected treatment technologies capable of reducing the level of arsenic in water below the standard limit (10μg/L) set by WHO are enlisted in Table 1 and 2.

In the later sections, discussions on the some of the most important techniques used for treatment of waterborne arsenic are presented. A number of these technologies find wide use in treatment of arsenic contaminated effluents in industries and while others offer solution to naturally affected areas through site remediation.

3.1 Preoxidation of As(III) in water

Almost in all commercial wastewaters and most natural waters the pH is generally below 9 which ensures most of the As(III) is in the form of H_3AsO_3. As there is no overall charge on H_3AsO_3, its removal through sorption and ion-exchange technologies are quite difficult [15,16]. However, in the same conditions, As(V) exists in the form of oxyanions $H_2AsO_4^-$ and $HAsO_4^{2-}$, where the negative charges on these species supplement the removal process. Hence, prior oxidation of waterborne As(III) into As(V) often boosts the removal process.

There are many reports of air and oxygen being too slow to oxidize As(III) into As(V) [11-13]. To be more precise, the half-life of As(III) in air and oxygen has been found out to be four to nine days and two to five days, respectively [13]. This has led the researchers to look out for more effective oxidizing methods. For example, when ozone (O_3) was passed through arsenic contaminated water, the half-life of As(III) drastically reduced to about four minutes [13]. Other methods used to speed up the oxidation in presence of air include use of sunlight, ultraviolet light, gamma radiation, electrochemical methods, chemical, and bacteria [11-38]. The oxidation of As(III) with ultraviolet radiation was observed to be significantly enhanced with the addition of iron(II) compounds, citrates and titanium dioxides (TiO_2) [19-20,25-29]. Several chemical oxidants have also been successfully used for the same purpose. Some of these noteworthy oxidants include chlorine (Cl_2), sodium hypochlorite (NaOCl), chlorine dioxide (ClO_2), potassium permanganate ($KMnO_4$), potassium ferrate (K_2FeO_4), cryptomelane (KMn_8O_{16}), manganite (MnOOH), pyrolusite (β-MnO_2), and manganese (Mn) oxide-coated sands [11,15,23,30-33]. H_2O_2 in the form of Fenton's reagent is also known to increase the oxidation rate of As(III) [17,34-35].

However, there are certain cases where the presence of interfering substances create problems during the oxidation of As(III). For instance, some naturally present organic substances such as, humic acids use up the oxidants themselves [39]. It has also been established that chlorine and ozone produce carcinogens upon reacting with organic matter and bromide ions, respectively [19, 33]. Moreover, when manganese left behind from the manganese-based oxidants exceeds the permissible drinking water standards, it may pose a serious threat itself [19]. Again, highly oxidizing yet relatively less toxic compounds such as potassium ferrate have proven to be expensive [23].

Hence, in the recent years the usage of such costly toxic oxidants has been discouraged and more emphasis has been laid on the cost effective choices of bacteria and microorganisms. Bacteria such as Microbacterium lacticum and Alcaligenes faecalis are known to oxidize As(III) [36-37]. Another worth mentioning methodology for oxidation of waterbone As(III) is the use of tartrate-citrate based photocatalyst extracted from the tamarind tree (Tamarindus indica) [38]. Such safe and cost effective techniques can prove very useful in developing countries such as India, Bangladesh, Nepal, etc.

3.2   Sorption

Sorption is referred to any process where adsorption, absorption, or both of them take place and the substances on which such processes take place are called sorbents. In most cases, to achieve best results with sorption, sorbents are suspended in a reactor. These sorbents are generally particles of diameter 0.3-0.6mm and include ion-exchange granules, fibres and other materials used for packing of chromatographic columns or filters. Due to their appropriate size, such sorbent particles provide sufficient surface area for efficient sorption of contaminants and allow the water molecules to flow through. Other than being inexpensive, such sorbents must be easily regenerated and must have high durability in presence of water. Some of the effective sorbent materials used for arsenic removal are discussed next and some sorption based materials which effectively reduce the arsenic contamination below 10µg/L are mentioned in Table 1.

3.2.1   Elemental iron

Iron in its elemental form is readily oxidized to rust in presence of water and oxygen. Rust contains a wide variety of iron compounds such as lepidocrocite (γ -FeO(OH)), sulfate ($4Fe(OH)_2 \cdot 2FeOOH \cdot FeSO_4 \cdot 4H_2O$), carbonate green rust, mackinawite (Fe_9S_8),magnetite (Fe_3O_4), maghemite (γ-Fe_2O_3), goethite (α-FeO(OH)), and amorphous ferrous sulfide (FeS). These compounds along with elemental iron are known to preferentially sorb and remove As(III)from water [3,41-44]. Nikolaidis et al. showed that iron fillings are quite effective in removing in As(III) with an initial concentration of about 300 µg/L from municipal waste leached water [43]. The columns packed with iron fillings were able to work efficiently for at least 8 months, after which significant leaching of iron starts. Another study showed that iron fillings, in addition to As(III) can also remove As(V) from aqueous solutions having arsenic concentration as high as 2 mg/L [44]. Moreover, it has been observed that the removal of As(III) and As(V) species from water becomes more efficient as the concentration of oxidized iron increases [45]. It must be noted that, elemental iron gets oxidized to iron oxides such as magnetite and removes As(III) and As(V) even when the water is devoid of molecular oxygen [45-46]. When iron nanospheres of diameter 1-120 nm were used, As(III) removal efficiency was observed to increase sufficiently [47]. At the same time, nano-alloys made up of elemental nickel-iron have also proved more effective for removal of As(V) [48].

Elemental iron has proved to be a simple, convenient and cost-effective sorption material for arsenic removal which can be utilized on a wide scale, especially in developing countries like Bangladesh [49].

3.2.2 (oxy)(hydr)oxides of iron, aluminium and manganese

The sorption of iron, manganese and aluminium(oxy)(hydr)oxides are largely due to the ion exchanges of OH_2^+, OH, and O^- groups present on their surface [50]. The amorphous (oxy)(hydr)oxides of these elements provide larger surface area compared to crystalline compounds and hence, the former are more suitable for arsenic removal through sorption [51].

Iron (oxy)(hydr)oxides are complexes consisting of hydrous oxides, hydrous hydroxides, and hydrous oxyhydroxides of iron(II) and iron(III), which can be easily obtained from rocks, sediments, or soils [52]. They are usually more effective in the sorption of As(V), rather than As(III) [53]. As(V) sorption takes place more effectively in acidic pH range (generally below pH 4). Under basic conditions, As(III) sorption relatively increases with the maximum As(III) sorption around pH 9. Similar to As(V), MMA and DMA sorption is more effective under acidic conditions [54]. Hematite ore based sorbents have been known to efficiently remove As(V) as high as 1 mg/L [55]. Goethite is also known to sorb As(V), whose efficiency can be further increased by doping with copper(II), nickel(II), or cobalt(II) [56-57]. Enhanced goethite based As(V) sorption was also observed when the ionic strength of the water was increased using potassium nitrate (KNO_3) [57]. Furthermore, sorption of As(V) can be improved by addition of lime which increases the amount of positive surface charges due to sorption of Ca^{2+} ion [58-59]. As(V) as well as As(III) removal through sorption is also known to increase with the use of cotton cellulose beads with iron (oxy)(hydr)oxides [60-61].

However, when the iron (oxy)(hydr)oxides granules are too coarse, they degenerate into impermeable sludges inside the columns used for water treatment [62]. Hence, to improve the stability and effectiveness of these compounds, they are often used as coatings on sands, cement granules or impregnated on activated carbon, ion-exchange fibers, mesoporous silica, mesoporous carbon, and zeolites [62-75]. In fact, iron (oxy)(hydr)oxide coated sands can effectively reduce As(III) as well as As(V) concentrations of 1 mg/L to below 10 μg/L [64-65]. Activated Bauxsol, a sorbent derived from Fe_2O_3 rich red mud has been observed to reduce the As(V) concentration below the standard limit as well [66]. ArsenX^{np} is a hybrid sorbent material made by impregnating the surface of polymeric beads with iron NPs and has proved very effective in reducing 50 μg/L of As(V) to below 10 μg/L [67]. Interferences from phosphates and silicates are known to cause problems during arsenic removal using iron (oxy)(hydr)oxides [60]. Iron(oxy)(hydr)oxide sorbents are generally easily rejuvenated by treating with sodium hydroxide (NaOH) and subsequently rinsing with sulfuric acid (H_2SO_4).

Manganese (oxy)(hydr)oxides are composed of pyrolusite (β-MnO_2), manganite (MnO(OH)), cryptomelane (KMn_8O_{16}), and birnessite [40,76]. Mine tailings and sea-bed rock concentrations are the chief sources of manganese (oxy)(hydr)oxides [40,76]. The iron-manganese tailings have high surface areas and porosities which may prove useful in arsenic removal [3]. Iron rich manganese ore have been successfully used to treat groundwater for removal of As(III) in the range 40-180 μg/L [77]. Mixed (oxy)(hydr)oxides of iron and manganese have proved very efficient in reducing100 μg/L of As(III) and As(V) inspiked tap water and mining effluents to below 10 μg/L [78]. Manganese (oxy)(hydr)oxides coated sands have been also used to treat water containing 500 μg/L As(III) and As(V) [79]. As(III) oxidation by manganese oxidizing bacteria, such as Leptothrix ochracea in manganese containing water is observed to significantly improve the arsenic removal efficiency [80].

Aluminium (oxy)(hydr)oxides include bayerite (α-Al(OH)$_3$, and gibbsite (γ -Al(OH)$_3$, diaspore (α-AlO(OH)) and boehmite (γ -AlO(OH)). Activated alumina used by most water treatment plants is generally generated by thermally dehydrating aluminum (oxy)(hydr)oxides to form amorphous, cubic(γ), and/or other polymorphs of corundum (Al_2O_3) [40,81-82]. However, activated alumina is not so proficient in removing As(III). Hence, As(III) is preoxidized to As(V) and subsequently filtered using columns of granular activated alumina [40]. Sorption of As(V) using alumina is found to be maximum in the pH range 5.5-6.0 [40]. Similar to the case of iron (oxy)(hydr)oxides, As(V) sorption on alumina is more effective under low pH conditions and decreases as the pH becomes alkaline.

Due to its reduced porosity, activated alumina often suffers from inefficient arsenic removal, which can be resolved by the use of mesoporous alumina [83]. A major problem associated with arsenic sorption on alumina is that the process slows down after twenty-four hours [25]. Activated alumina is generally regenerated or disposed after one to three months [33]. Regeneration of alumina includes washing with NaOH, followed by rinsing with sulphuric or hydrochloric acid. However, disposal being inexpensive is preferred over regeneration as regeneration often discharges harmful effluents. Moreover, even after regeneration some amount of arsenic is left retained on alumina [25]. Interference from chloride, fluoride, phosphate, silica, organic materials, and sulfate are additional disadvantages related to As(V) sorption on activated alumina [33,40,84].

Improved arsenic removal have been achieved by combining aluminium compounds with manganese or/and iron [40,85-87]. For example, AlcanAAFS-50 is an iron-doped alumina sorbent where the arsenic sorption has been observed to betwo to five times of the normal capacity of ordinary alumina and is less sensitive to pH variations [40]. Calcinated bauxite based sorbents have proved effective in treating water containing 2 mg/L of As(V) [88]. Another worth mentioning sorbent is Hydrotalcite ($Mg_6Al_2(CO_3)(OH)_{16} \cdot 4H_2O$), which removes As(III) as well as As(V) through sorption and/or exchanges with interlayer anions [89-90].

3.2.3 Titanium dioxides(TiO₂)

Titanium dioxidealso removes As(III) and As(V) from water via sorption. [19,91-95]. Anatase (a mineral of TiO_2), in its granular form is known to reduce 300 µg/L As(III) and As(V) in spiked groundwater samples to below 10 µg/L [91]. TiO_2 based nanoparticles (NPs) have been also observed to remove MMA and DMA from water [92]. Manna et al showed that crystals of hydrous titanium dioxide can quickly remove significant amounts of arsenic, even in neutral water having pH 7 [94].

3.2.4 Activated and impregnated activated carbon

Activated carbon is amorphous and highly porous graphite which has large surface area as well. It can be easily produced from wood, coal, crop residues, bones and petroleum [82]. Although, activated charcoal is used for removal of many impurities from air as well as water, it does not possess the same effectiveness for arsenic removal [84-85]. High cost and regeneration issues also make the situation complicated [86]. Moreover, the use of finely powdered form of activated carbon produces impregnable sludge in water.

To overcome such issues, granular activated carbon impregnated with iron, calcium chloride ($CaCl_2$), copper, and zirconium salts, have been employed [82,96,99-101]. Iron impregnated granular activated carbon has been observed to remove As(V) from spiked ground water samples [102]. Polyanilines (($C_6H_7N_4)_n$) and other organic compounds such as L-cysteine methyl ester ($CH_3OOCCH(CH_2SH)NH_2$), when used along with activated carbon are also know to significantly improve its arsenic removal efficiency [103-104]. However, at pH values greater than 8.5, considerable interference from phosphate and silicate ions have been during the removal of As(V) using L-cysteine methyl ester [104].

Materials Research Forum LLC
https://doi.org/10.21741/9781644901144-4

3.2.5 Biomass

Sorghum and rice husk wastes have proven to be efficient as well as cost effective materials for treating As(V) in water [105-106]. Their arsenic removal ability was further enhanced upon addition of iron salts such as $FeSO_4$ or $Fe(NO_3)_2$. Sulfate ions and humic acids severely interfere with sorption of As(V) on sorghum biomass. More crop and crop waste based arsenic removal methods are mentioned under the bioremediation section discussed later. As(III) removal was also achieved using ammonium thioglycolate treated ground chicken feathers [107].

3.2.6 Ion-exchange resins and metal-loaded gels

A wide range ion-exchange resins, polymers and metal-loaded gels have been discovered for arsenic removal from water [67,84,108-115]. Such ion-exchange resins are made up of mostly natural or synthetic organic polymers and some inorganic substances. Most of the gels and polymers can be packed inside a column while some of the polymers themselves sorb arsenic [108]. Moreover, there are cases, where polymers and gels only act as support media for other materials meant for sorption [66,109-110]. Similar to most of the sorption materials, these resins are generally more efficient for As(V) removal compared to As(III). In case of arsenic removal, weak-base resins are known to work only in a narrow pH range. Whereas strong-base resins which contain amine and quaternary ammonium (NR_4^+), groups are more effective in removal As(V) over a pH range of 7-9 [40,84]. For instance, Amberlite IRA-458,is known to remove about 90% of 1 mg/L of As(V) in just five minutes [84].

Advantages of ion-exchange resins include insensitivity to water flow variations and easy regeneration. On the other hand, the demerits of using ion-exchange resins include large volume of disposables after regeneration, variable removal rates in presence of interfering species, incompetence in case of high amount of total dissolved solids (TDS), and interferences from chloride, sulfate, organic materials and other chemicals [40]. Moreover, as ion-exchange resins become saturated after sorption, the already sorbed arsenic may start leaking, which is known as chromatographic peaking. However, this can be easily evaded by changing the resins before the saturation sets in [40].

To increase the efficiency of the polymers and gels, they are often impregnated with cations. These substances are known as metal loaded gels and are prepared by passing or mixing aqueous solutions of the metal ions solutions over the polymers or gels. The commonly used metals ions are iron(III), aluminium(III), copper(II), zirconium(IV), manganese(IV),lanthanum(III), titanium(IV), and molybdenum(VI) [67,84,109-116]. Strong base ion-exchange resins like Purolite A-505 and Relite A-490 along with zirconium loaded anion-exchange resins are known to remove sub mg/L As(V) from laboratory solutions [115-116]. Hydrated form of stannic oxide (SnO_4) has proved to be an exceptional sorbent material by reducing As(III) and As(V) concentrations from several mg/L to below 10 µg/L [117]. All metal-loaded polymers, except iron, are less affected by chloride and sulfate interferences [84]. A couple of highly effective and long serving arsenic sorption materials developed by the dispersion of iron (oxy)(hydr)oxide NPs over a polymer base are already mentioned previously [51,67]. Resins loaded with zirconium were found to be efficient for As(III) sorption [84,116]. Fluoride and phosphate ions are known to cause interference in the zirconium loaded resins [75,116].

It can be summarized that the use of ion exchange resins and metal loaded gels have proved to be one of the most effective methods of arsenic removal from water. However, problems such as regeneration of the resin, disposal of used up resins and wastes remain as issues to be addressed.

Table 1. Selected sorption based waterborne arsenic removal methods capable of reducing the arsenic contamination below the standard limit set by WHO (10 µg/L).

Treatment Method	Sample type	As species	Injected As concentration for treatment [mg/L].	Ref.
Elemental iron sorbent	Municipal landfill leachate	As(III)	0.3	[43]
	Laboratory Solutions	As(III)/As(V)	2.0	[44]
Hematite ore deposit sorbents	Laboratory Solutions	As(V)	1.0	[55]
Fe-loaded anion-exchange polymer	Laboratory solutions	As(III)	0.1	[51]
		As(V)	0.1	
		As(V)	0.1	[63]
Fe (oxy)(hydr)oxide-coated sand filtration	Spiked ground water	As(III)/AsV)	1.0	[64]
		As(III)/AsV)	1.0	[65]
Activated Bauxsol sorbent	Laboratory solutions	As(V)	0.05-16.6	[66]

ArsenX^{np} sorbent	Ground water	As(V)	0.011-0.023	[67]
Hardened Portland cement sorbent	Laboratory solutions	As(V)	0.2	[68]
Ion-exchange polymeric fibers impregnated with Fe-(oxy)(hydr)oxide	Laboratory solutions	As(III)	500	[69]
		As(V)	500	
Fe-rich Mn ore sorbent	Ground water	As(III)	0.04-0.18	[77]
Fe and Mn (oxy)(hydr)oxide sorbents	Laboratory Solutions	As(III)	0.100	[78]
		As(V)	0.100	
Mn (oxy)(hydr)oxide-coated sand filtration	Laboratory Solutions	As(III)	1.0-0.5	[79]
		As(V)	0.5	
Mn (oxy)(hydr)oxide sorbents + Mn-oxidizing bacteria	Laboratory Solutions	As(III)	0.035	[80]
		As(V)	0.042	
Calcinated bauxite sorbent	Laboratory Solutions	As(V)	2.0	[88].
Hydrotalcite sorbent	Laboratory Solutions	As(III)	0.4	[89]
TiO$_2$ (granular anatase) sorbents	Spiked ground water	As(III)	0.3	[91]
		As(V)	0.3	
Fe(III)-impregnated granular activated carbon sorbent	Spiked ground water	As(V)	1.031	[102]
Polyaniline in activated carbon sorbent	Laboratory Solutions	As(V)	0.15 and 8.0	[103]
L-cysteine methyl ester modified carbon powder sorbent	Laboratory Solutions	As(III)	0.070	[104]
Hydrated SnO$_4$ sorbent	Laboratory Solutions	As(III)	5.0	[117]

3.3 Precipitation/coprecipitation

Arsenic removal from water through precipitation involves addition of certain specific chemicals such as iron or aluminium salts, which have the ability to convert the dissolved arsenic into insoluble solids and separate out. Coprecipitation is a kind of precipitation technique where the substance to be removed (such as arsenic) sorbs on the surface or bulk of other precipitates. In the next step, these precipitates are removed from water by filtration, flotation, sedimentation and centrifugation. At last, thickening and dewatering processes are carried out to decrease the amount of sludge, which reduces the transportation and disposal costs. However, the dewatered solids must undergo regulatory tests before disposal. These methods are comparatively cost effective and efficient for treating water with high TDS. Moreover, these methods may prove more effective in large-scale water treatment plants. Some of the common precipitation/coprecipitation chemicals are discussed below and some of the selected effective precipitation/coprecipitation methods are enlisted in Table 2 along with other methods.

3.3.1 Lime (CaO)

In the past, addition of lime (CaO) to contaminated water has been used numerous times for As(V) removal [33,118]. Lime has been found to precipitate almost 90% of As(V) when the pH values are maintained at 10.5 or higher. However, it has proved less efficient for As(III) precipitation [33]. Moreover, stability of the calcium arsenate precipitates may lead to disposal problems [119]. Furthermore, acidic conditions, presences of carbonates, bicarbonates, or carbon dioxide can decompose calcium arsenates, resulting in release of arsenic in the ecosystem [120-121].

3.3.2 Iron and aluminium salts

Iron (III) salts, such as ferric chloride and ferric sulfate, have proved efficient for As(V) removal from water [39]. After peroxidation, As(III) can also be removed using iron(III) salts [33]. Several dosages of iron (II) over a few hours aids in the peroxidation of As(III) [122]. Addition of a mixture potassium hydrogen phospahte (KH_2PO_4) and hydrated sodium sillicate ($Na_2SiO_3.9H_2O$) can improve arsenic removal [122]. Regular addition of small amounts of aluminum sulfate along with iron(III) salts are also known to make arsenic removal from water more effective [123]. Fe-(oxy)(hydr)oxide based coprecipitation followed by filtration have proved efficient in removing up to 50 µg/L of As(V) from real and spiked water samples [124-125]. Siderite based coprecipitation methods have been tested with laboratory prepared solutions and have shown excellent abilities of removing arsenic up to 2 mg/L concentrations [126]. However, interferences from natural organic matter, phosphate or silicate may cause problems in the use of iron (III) salts [25,124,125]. Again, ferric sulphate is known to reduce the life of the water treatment tanks due to corrosion [39,127]. In this regard, sacrificial electrochemical coprecipitation with iron electrodes can result inbetter coprecipitation of As(V). For, instance, an electrochemical unit using $Fe(OH)_3$ based electrochemical coprecipitation with sacrificial anode was developed for treating household water by Sagitova et al [128]. Another electrochemical coprecipitation method with iron and zinc electrodes also reports of efficiently treating 0.5 mg/L of arsenic [21]. Regular filtration of the precipitates using membranes and sand can further improve the precipitation efficiency. This can prove to be an inexpensive and efficient method for As(V) removal from household water supplies [125].

Less effective aluminium sulfate ($Al_2(SO_4)_3$) has also been used for coprecipitation of As(V) [25]. Coprecipitation with aluminium (oxy)(hydr)oxides followed by filtration has shown that it can reduce arsenic concentrations to below 10 µg/L [124]. However, the coprecipitation of arsenic is hindered by the fact that aluminium hydroxides are soluble in water, which would release the arsenic precipitates [124].

Table 2. Non-sorption arsenic removal methods which have proved effective in reducing arsenic contamination below the 10 µg/L limit.

Treatment Method	Sample type	As species	Injected As concentration for treatment [mg/L].	Ref.
Fe-(oxy)(hydr)oxide coprecipitation followed by filtration	Spiked source/ artificial water	As(V)	0.004-0.052	[124]
	Blended well water	As(V)	<0.050	[125]
Siderite based coprecipitation (with possible sorption)	Laboratory solutions	As(III)	0.250–2.000	[126]
		As(V)	0.250–2.000	
Fe(OH)₃ based electrochemical coprecipitation with sacrificial anode	Laboratory Solutions	As(III) oxidized to As(V)	0.500	[128]
Electrochemical coprecipitation with Fe and Zn electrodes followed by filtration	Laboratory solutions	As(V)	0.070, 0.100 and 0.130	[21]
	Synthetic water	As(III) (preoxidized)	0.5	
Al-(oxy)(hydr)oxide coprecipitation followed by filtration	Spiked source/ artificial water	As(V)	0.004-0.052	[124]

PRBs with lime, iron oxides, and limestone	Laboratory solutions	1:1 mixture of As(V) + As(III)	1000	[129]
Cationic surfactant and ultrafiltration	Drinking water	As(V)	0.05-0.250	[133]
Nanofiltration	Synthetic water	As(V)	0.1	[137]
	surface water	As(V)	0.100-0.382	
Ce-iron oxide based magnetic sorbent	Laboratory Solutions	As(V)	1.0	[139]
Water hyacinth dried roots sorbent	Spiked tap water	As(V)	0.200	[145]
Maple Wood ash sorbent	Laboratory Solutions	As(III)	0.1-0.5	[146]
		As(V)	0.25-0.5	
Coconut coir pith anion exchanger	Synthetic water	As(V)	1.0	[147]
Bioaccumulation with Fern (*Pteris sp.*)	Laboratory Solutions	As(V)	0.020-0.200	[155]
Bioaccumulation with modified bacteria	Spiked contaminated water	As(III)	0.05	[163]

3.4 Permeable reactive barriers (PRBs)

Permeable reactive barriers (PRBs) are membranes or barriers which allow selective passage of some materials while blocking others. PRBs have proved very useful in treating arsenic contaminated water having very high levels of arsenic [129]. As soon as the arsenic contaminated water interacts with these barriers, arsenic removal gets initiated through sorption, ion-exchange, biodegradation, precipitation/coprecipitation, or filtration. Most, PRBs use zero valent iron and occasionally lime, portlandite, steel industry by-products are also used as the barrier material [3,42]. PRBs are very efficient for arsenic removal as well as other contaminants [129]. However, there is much work is still needed to be done regarding designing, installation, long-term monitoring, possible recovery and proper waste disposal of these permeable reactive barriers.

3.5 Filtration and membrane techniques

Filtration of contaminated water is generally done under pressure so that the contaminated water is forced through a selective semipermeable membrane and this process is known as pressure filtration. There are four kinds of pressure filtration, namely, microfiltration, ultrafiltration, nanofiltration and reverse osmosis [98,130-137]. *Microfiltration* is useful when the diameter of the contaminated particles is larger than 0.1 μm. Only larger arsenic precipitates can be removed using microfiltration after precipitation/coprecipitation with iron(III) salts [98,130]. However, the relatively larger grain size of microfilters proves difficult to separate dissolved and colloidal arsenic.

Ultrafiltration is performed for removing particles as small as 0.01 μm. Similar to microfilters, the grain size of ultrafilters renders them incapable of removing dissolved arsenic [98]. As(V) removal using ultrafilters can be improved by electrodialysis and presence of dissolved organic carbon [98,132]. Cationic surfactant aided ultrafiltration methods have proved useful in treating drinking water having a As(V) contamination in the range 50-250 μg/L [133].

Nanofiltration separates particles whose diameter are in the nanometer dimensions and involves the principles of charge and size exclusion [98,135]. Hence, the ability of a certain nanofilter is decided by its electrical charge and permeability. As a result, inorganic As(V) is more efficiently removed than As(III) [98,136]. The removal of As(III) and DMA(V) with nanofilters are known to improve at relatively higher pH values. Moreover, arsenic removal efficiency with nanofilters is also observed to improve at higher pressures [98]. Sometimes, the negative charges on the surface of nanofilters repel the chemical species carrying similar charges such as oxyanions of arsenic and thus, aiding their removal. This process is known as Donnan exclusion and often occurs simultaneous along with nanofiltration [136]. Nanofiltration has been successfully utilized to treat surface waters having 100-382 μg/L of As(V) [137].

Reverse osmosis is known to filter very small particles having a diameter of approximately 0.0001 μm. Reverse osmosis is well known to remove As(V) efficiently (up to 98%), by utilizing membranes generally made up of cellulose acetate, polyamides, polyvinyl alcohol, and other synthetic materials [98]. However, even at high pH, effectiveness of As(III) removal remains relatively low at around 48-75% [33]. Peroxidation of As(III) does not offer any help in this case as the oxidants may damage the membrane [98].

Generally, the processes of nanofiltration and reverse osmosis involve capillary flow and solution diffusion, respectively, and both need high pressures leading to slightly higher energy consumption and expenses. On the other hand, comparatively less efficient microfiltration and ultrafiltration techniques which remove contaminants through mechanical sieving require relatively lower pressures and are inexpensive.

3.6   Magnetic treatment

Magnetic properties of magnetite (Fe_3O_4) along with its high surface area have been utilized to develop good sorbent material for As(V) and As(III)in the form of NPs. When homogenously distributed in contaminated water, magnetite NPs having a diameter of 12 nm show increased efficiency for arsenic removal [138]. When cerium(IV) are used as dopants in magnetite, As(V) sorption increases significantly and reduces arsenic contamination below 10 µg/L [139]. Again, compared to natural magnetite, synthetically prepared magnetite has proved more efficient for arsenic sorption, especially when coated with calcium alginate [140]. Magnetite has been also used along with activated charcoal sorbents for arsenic removal [141]. Dopinghydrotalcite with excess iron(III) and nickel(II) also increases magnetic sorption based arsenic recovery from clays [142]. Superconducting magnets have been also utilized for arsenic removal [143]. Although superconducting magnets were able to reduce the arsenic concentration considerably, the level of arsenic left in the water after the treatment was over the WHO recommended limit of 10 µg/L. Moreover, the use of superconducting magnets may prove too demanding for developing countries.

3.7   Bioremediation and biological treatment

Living organisms or substances derived from them are also capable of sorption of arsenic. Crop produce, crop wastes, ferns, algae, and chitosan have been observed to remove arsenic from water [120,144-159]. Bacteria, fungi, plants, and other microorganisms are also known to remove arsenic from water [160-165].

Some plants, plant parts or their wastes are capable of accumulating arsenic biologically from the environment. This process is generally known as phytoremediation. For example, sorption of arsenic iron coatings present in the roots of the rice plant (Oryza sativa L.) leads to arsenic accumulation in the plant which is well known [144]. Dried roots of water hyacinth, ash derived from maple wood and coconut coir have proved very effective in lowering arsenic concentrations to below 10 µg/L [145-147]. Even simple aquatic plants and ferns such as, Apium nodiflorum, Salvinia natans and Pteris vittata have been observed to sorb As(V) [148-155]. It has also been found that there is preferential bioaccumulation of arsenic in the stems and leaves of Pteris cretica cv Mayii [153]. Furthermore, high levels of arsenic were discovered in old needles of Pinus pinaster, Calluna vulgaris, and C. tridentatum and in leaves from C. ladanifer, Erica umbellate, and Quercus ilex subsp. ballota [156]. Such kind of phytoremediation can be additionally improved with the aid genetic engineering [157]. Even after the cost-effectiveness of such phytoremediation, large scale implementation of the process suffers from slowness and may be stalled by drought and pests. Over a long period of time, changes in chemical and physical conditions of a site could also render phytoremediation ineffective. Moreover, disposal of arsenic enriched biomass could prove to be an additional issue.

Bacteria such as Gallionella ferruginea and L. ochracea when incorporated in fixed-bed up-flow bioreactors are known to treat arsenic-spiked groundwater by oxidizing As(III) to As(V) and subsequent removal by precipation/coprecipitation [161-162]. Some bacteria, such as Escherichia coli and Acidithiobacillus ferrooxidans directly removeAs(III) from water [163]. Moreover, colonies of a bacteria named, Gallionella sp was observed to induce coprecipitation of As(V) in iron-rich mine water [164]. The methylated sludge obtained after the chemical modification of dead fungal (P. chrysogenum) biomasses are also known for effective sorption of As(V) from wastewaters [165]. Thus, bacteria and fungi provide us with cheaper and safer methods of arsenic removal from water. However, such methods should be very carefully monitored and managed to prevent significant methylation of inorganic arsenic which often leads to formation of highly hazardous methylarsine gases.

Many naturally occurring process biodegrade, precipitate, sorb, as well as reduce the toxicity and mobility of arsenicwithout any significant human activity. For example, sorption of arsenic in iron rich ores under oxidizing conditions and coprecipitation with sulfides under strongly reducing conditions [166]. Although it offers a smart and inexpensive solution, this method is too slow to meet the immediate concerns.

4. Analysis of arsenic contaminated water

As already mentioned in the previous sections, arsenic contamination in water can prove to be very harmful for the human population as well as the overall ecosystem. Even though the content of arsenic in ambient air is in the range of ng/m^3 which is considered relatively low, its prolonged exposure may cause health issues for workers involved in arsenic related industries [4-5]. Briefly, emission of arsenic in the form of particulate matter or volatile arsine gas can prove a serious health hazard for the workers in the smelting, coal combustion and semiconductor industry. Disposal without efficient treatment or improper disposal of arsenic contaminated solid wastes and water discharges can potentially pollute soils, sediments, and safe groundwater sources. Nevertheless, issues due consumption of arsenic contaminated water has taken enormous proportions all over the world. Hence, testing of arsenic in industrial effluents, natural aquifers as well as drinking water resources of the affected areas has become ubiquitous. Moreover, quantitative estimation of arsenic in water is also essential for the correct selection of the preferred treatment methodology.

Materials Research Forum LLC
https://doi.org/10.21741/9781644901144-4

Several analytical techniques are used for arsenic estimation which include colorimetric, spectrophotometric, electrochemical, atomic absorption and mass spectrometric methods [167-267]. In addition to these state-of the art expensive methods, recently many cost effective field techniques have been also developed. Although, the sample preparation methods greatly differ in case of contaminated water, solids and gases, the quantifying techniques used for arsenic are more or less common in each case. In the later sections, the different analytical techniques for estimation of the arsenic are discussed. Considering the fact that globally millions of people are affected due to drinking arsenic contaminated water, some of the highly sensitive arsenic estimation techniques which might find use in large scale field analysis are also summarized in Table 3 and Table 4.

4.1 Techniques analysis of waterborne arsenic

4.1.1 Colorimetric methods

Inexpensive nature combined with relatively less complications of sample preparations, have made several colorimetry based methods ideal for field tests. Some of the highly sensitive colorimetric methods used for arsenic quantification are presented in Table 3. Based on these techniques, a number of color based test kits for arsenic have been developed over the years [167-171]. Almost all of these colorimetric tests are based on the Gutzeit method which was first proposed in the year 1879 [172]. In the Gutzeit method, inorganic arsenic species containing As(V) and As(III) are converted into volatile arsine (AsH$_3$) gas using a suitable reducing agent such as a combination of zinc and hydrochloric acid. The arsine gas is subsequently made to react with silver nitrate (AgNO$_3$) to form dark coloured complex. The colour of the complex varies from light grey to black depending upon the amount of arsenic in the sample. Later, expensive silver nitrate crystals were replaced by a strip of filter paper impregnated with mercuric bromide which formed a yellow stain upon reaction with arsine [173]. The amount of the arsenic is generally estimated by comparing the colour of the crystal or the stain with a standard chart. Initially, the Gutzeit method could estimate arsenic with the lower limit of detection being 100 µg/L. However, some of these automated kits claim to detect arsenic as low as 10 µg/L [167-171]. The Gutzeit method can also be utilized to speciate As(III) and As(V). Conversion of As(III) species into arsine can be achieved under mild acidic conditions (at pH ~ 4), whereas for conversion of As(V) into arsine, the pH should be 1 or less.

Other than the limit of detection being high, Gutziet methods also suffer from a number of drawbacks including use of toxic mercuric bromide, leakage of harmful arsine gas, color degradation in sunlight and human error in colour identification resulting in inaccuracy and irreproducible results. Moreover, presence of arsenic in zinc (used for conversion to AsH_3) as an impurity, can result in ambiguous estimations. This method with slight modifications has been widely used for developing countries like Bangladesh, where the concentration of arsenic in groundwater often exceeds 100 µg/L [174-177]. Lately, there are reports that these test kits have proved inaccurate [178-179]. With the use of automated delivery and detection techniques many of the human error related problems can be negated to yield better accuracy and reproducibility. Moreover, the lower detection limit of the Gutzeit method can be further improved by using spectrophotometers for measuring the color intensity. For instance, with the aid of image scanning and computation, Salman et al. were able to achieve a limit of detection of 1 µg/L for As(III) analysis in well and canal water [180]. Nevertheless, automation and the use of spectrophotometers increase the cost per test as well as require skilled manpower.

Another important colorimetric method for arsenic is the molybdenum blue test [181-183]. In this method, As(V) reacts with molybdate forming a molybdenum blue complex, whose color intensity is measured to estimate the amount of As(V). As(III) must be peroxidised to As(V) in order to estimate the total inorganic arsenic. However, this method is less sensitive and often suffers from interferences from silica and iron (III). Later, addition of a preconcentration step significantly improved the accuracy of the method (limit of detection of about 0.08 µg/L) [184]. In another method, the generated arsine gas is made to react with silver diethyl dithiocarbamate dissolved in pyridine which results in a red colored complex [185-186]. The intensity of the red colored complex was measured spectrophotometrically at 536 nm, to estimate the amount of arsenic. Silver diethyl dithiocarbamate forms colored complexes with many metals and thus,the number of interfering sources also increases [187]. Moreover, the molybdenum blue and silver diethyl dithiocarbamate methods involve the use of harmful chemicals such as potassium bromate ($KBrO_3$) and pyridine, respectively. Another colorimetric method utilized the adsorption of a cationic dye named ethylviolet on molybodoarsenate to form ionic associates in the form of particulate matter and achieve a limit of detection of 4 µg/L [188]. However, the process is lengthy and arduous.

Color based detection of arsenic using functionalized gold (Au) NPs have also proved to be very sensitive [189-190]. Moreover, interferences from other metal ions are almost absent in such studies. Specifically, whenever arsenic binds to these functionalized AuNPs, they develop a distinct colour. Again, the intensity of the color measured spectrophotometrically at a suitable wavelength gives an estimate of the arsenic present in the sample. With detection limits of sub µg/L level being achieved in these studies, such methods can be very handy in detection of trace levels of arsenic in potable drinking water. However, use of expensive reagents and spectrophotometers demand a lot of capital and skilled labour.

4.1.2 Other spectroscopic methods

With the advancement of technology, several spectroscopy based techniques have been explored for arsenic measurement. These methods include Laser-induced breakdown spectroscopy (LIBS), Surface-enhanced Raman spectroscopy (SERS), Total reflection X-ray fluorescence spectrometry (TXRF), Attenuated total reflectance Fourier transform infrared spectroscopy (ATR-FTIR), Surface plasmon resonance (SPR) based sensors [191-199].

LIBS has been proved to be a very quick analytical tool for arsenic determination in wastewater [191]. SERS is based on the detection of surface plasmon activity shown by the contaminants when adsorbed on certain surfaces such asAu, Ag and CuNPs [192-193]. At the same time, TXRF has proved very slow while arsenic estimation in seawater tailings, from gold-copper mine, acid mine drainage water [194-197]. ATR-FTIR measurements have very poor sensitivity towards waterborne arsenic [198]. SPR sensors using thiol based organic compounds as coatings have been claimed to detect arsenic as low as 1.5 ng/L [199]. However, these results fail to match with graphite furnace atomic absorption spectroscopy (GF-AAS) data and with real samples SPR sensors are known to produce ambiguous results [199-200].

Most these methods involve the use of the bulky, expensive instruments which require high maintenance cost and skilled manpower. Moreover, with the limit of detection mostly being in the range g/L-mg/L, the sensitivity towards arsenic measurement needs to be improved vastly for these methods.

4.1.3 Luminescence-based methods

Arsenic estimation by measuring the fluorescence produced upon reaction of As(III) or As(V) with surface derivatized quantum dots (QDs) such as CdS, L-cysteine capped CdS, and glutathione (GSH)-capped CdTe QDs have also been probed [201-203]. Although, a limit of detection of 0.75 µg/L is encouraging, cost-effective methods of detecting fluorescence must be explored to achieve arsenic analysis on a large scale. Liquid phase chemiluminescence (LPCL) detection of As(V) with a limit of detection of 0.15 µg/L has proved to be more sensitive and cost-effective comparatively [204-205]. It involves reaction of As(V) with molybdate in acidified vanadate solution, resulting in formation of vanadomolybdoarsenic heteropoly acid (V-HPA). In the next step, Subsequently, V-HPA complex is reacted with alkaline solution of luminol to produce a signal which is measured by a photo multiplier tube (PMT). However, phosphate and other heavy metals must be co-precipitated to remove any interference.

Table 3. Highly sensitive colorimetry and luminescence based arsenic analysis techniques with field deployable abilities.

Method	Sample type	Arsenic Species	Limit of detection [μg/L].	Practical range [μg/L].	Ref.
HgBr$_2$ impregnated paper	Well/canal water	As(III)	1	2-20	[180]
Molybdenum Blue	Groundwater	total As	8	10-10,000	[183]
Molybdenum Blue	River/well/ lake water	As(III)/ total As	0.07/0.09	0-50	[184]
Molybdoarsenate-ethylviolet complex	Tap/river/ ground/mine water	As(V)	4	0-300	[188]
Spectrophotometric using Au-NPs	Well/tap/ bottled water	total As	0.003	0.001-50	[189]
Spectrophotometric using Au-NPs	Laboratory solutions	As(III)	0.6	0-100	[190]
Mg-Coprecipitaion + VMA-HPA Luminol, LPCL	Fresh water	As(V)	0.15	0.15-7	[204]
VMA-HPA preconditioned Luminol, LPCL	Mineral/ tap drinking water	As(V)	6.7	7.5-37,500	[205]
HG- gas diffusion - O$_3$ -GPCL	Drinking water	As(III)	0.6	0.6-25,000	[207]
Electrochemical HG- O$_3$- GPCL	Tap/ Spiked water	Total As	0.36	-	[208]
Electrochemical HG-O$_3$-GPCL	Ground/ tap water	As(III)/ total As	0.76/ 0.09	-	[209]
HG-O$_3$-GPCL	Ground/ tap water	As(III)/ As(V)	0.05/0.09	-	[210]
Manual HG-O$_3$- GPCL	Ground/ tap water	Total As	1.0	-	[212]

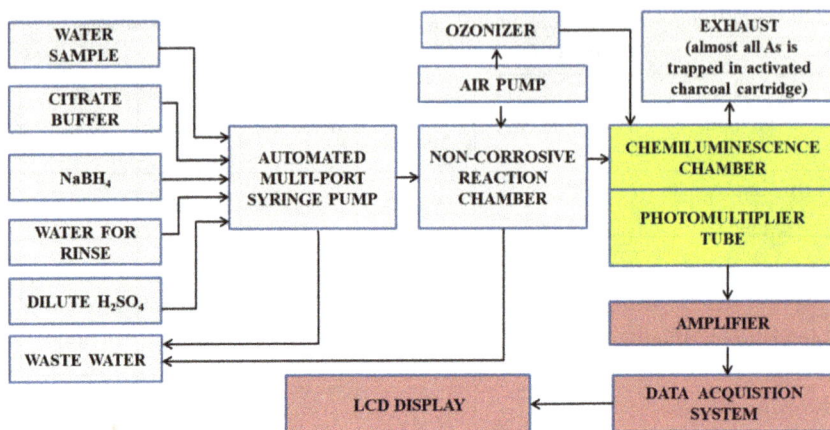

Figure 1. A schematic diagram of automated gas phase chemiluminescence based arsenic detection system.

The gas phase chemiluminescence(GPCL) technique is based on the measurement of chemiluminescence produced during the reaction between arsine and ozone, initially reported by Fujiwara et al [206]. Utilizing this GPCL, several studies have reported cost-effective methods of detecting sub μg/L level of inorganic arsenic in water [207-212]. For instance, Idowu et al. [210]. developed a fully automated instrument to demonstrate sequential analysis As(III) and As(V) along with determination of total inorganic arsenic. No significant interferences were observed from phosphates and other heavy metals. Later, efforts were made to this instrument small, compact, portable and robust [211]. A schematic diagram of the automated GPCL based detection system is presented in Figure 1. The detailed description of the same is already available in the literature [210-211]. A manually operated system was also explored to make it more affordable for countries like Bangladesh, where manual labour is relatively reasonable [212]. These GPCL based methods are highly sensitive with the limit of detection mostly being lower than 1 μg/L. In addition to low cost per test, automation opens the choice of untrained labour which could prove very effective solution for waterborne arsenic analysis in some of the developed countries as well. Some of the field deployable highly sensitive luminescence based arsenic detection techniques are mentioned in Table 3.

Advances in Wastewater Treatment I   Materials Research Forum LLC
Materials Research Foundations **91** (2021) 111-170   https://doi.org/10.21741/9781644901144-4

4.1.4 Hydride generation -atomic absorption spectroscopy (HG-AAS)

Atomic Absorption Spectroscopy (AAS) is a well-established instrumental technique where minute amount of metals contaminants in parts per million (ppm) or parts per billion (ppb) level concentrations can be estimated accurately and reproducibly [213]. In this method, the metal contaminants are volatilized, atomized and burnt in high temperature flame. The amount of the metal contaminant is estimated by measuring the intensity of the spectroscopic signatures of the metal using an absorption spectrophotometer.

Arsenic metal itself is not volatile and hence, to convert all the arsenic into volatile arsenic hydride (arsine gas) quantitatively, reduction with sodium borohydride ($NaBH_4$) is performed [214-216]. In this method, hydride generation (HG) is much efficient and the issue of erroneous measurement due to the use of zinc or magnesium is resolved. Specifically, for arsenic determination using HG-AAS, the sample is acidified and excess $NaBH_4$ is added to quantitatively convert all the different forms of arsenic into volatile arsine gas. The resulting arsine gas is then carried by an inert gas such as nitrogen or argon to the atomizer. Arsine is atomized by heating it inside a quartz tube by an air/acetylene flame. Finally, the amount of arsenic in the flame is estimated by measuring the intensity of a characteristic wavelength of 193.7 nm using a spectrophotometer. This HG method significantly enhances the sensitivity of detecting arsenic using AAS and can also be used with other detection techniques [217]. Again, in HG-AAS, interferences due to overlapping spectrum are greatly reduced [218]. However, HG-AAS alone cannot speciate between As(III) and As(V),that is, only the total arsenic content can be determined. This issue can be resolved by coupling HG-AAS with one of the many chromatography based speciation techniques [219]. Moreover, due to requirement of skilled labour combined with high cost per test, and maintenance cost HG-AAS is not recommended for large scale arsenic measurements in the field.

4.1.5 Electrothermal atomic absorption spectroscopy (ETAAS)

In this AAS method, samples are deposited in a small graphite coated tube which can then be heated using high electric current to vaporize and atomize the analyte [213]. Hence it is also known as graphite furnace atomic absorption spectroscopy (GF-AAS). Chemical matrix modifiers are also introduced along with the samples to stabilize the analyte and increase the volatility of the sample matrix. The most widely used chemical modifier for arsenic estimation is $Pd(NO_3)_2$ [219-222]. Au and AgNPs are also reported to serve as good chemical modifiers for reducing any interferences arising in sea water or water rich in sodium ions [223-224]. These substances aid in increasing the pyrolysis temperature beyond 1100 °C, thus increasing the sensitivity of arsenic detection. Other modifiers such as a combination of zirconium(Zr) with iridium (Ir) and mixture of tungsten with noble metals(such as W-Rh, W-Ru, W-Ir) have proved very efficient for arsenic estimation in sludge, soil, sediments, coal, ashes and water by ETAAS [225-226]. Similar to HG-AAS, ETAAS must be combined with a chromatographic separation technique for arsenic speciation [218].

4.1.6 Electrochemical methods

Electrochemical methods such as cyclic voltammetry (CV), cathodic stripping voltammetry (CSV) and anodic stripping voltammetry (ASV) using a variety of electrodes have been explored for arsenic analysis [227-242]. Most of such electrochemical methods are cost effective and at the same time, they have proved to be immensely sensitive. Some of the most sensitive electrochemical techniques are enlisted in Table 4. Specifically, the limit of detection was observed to be 0.3 μg/L in a ASV study of total arsenic analysis using a coating of ammonium 2-amino-1-cyclopentene-1-dithiocarboxylate (AACD) on a mercury (Hg) film electrode with prior Cr(III) accumulation [229]. It is noteworthy to mention the work of Gibbon-Wash et al where ASV studies using a Mn coated Au microwire presented a limit of detection of 0.15 ng/L [231]. Differential pulse anode stripping voltammetry (DPASV) using vibrating Au microwire and Au surface have also shown sub μg/L detection limits [232-233]. ASV on glassy carbon electrode (GCE) modified with coatings of Au/Pd NPsand microwalled carbon nanotubes (MWCNT) have also proved sensitive for arsenic analysis [234-235]. Anode stripping linear sweep voltammetry (ASLSV) studies with mercaptoethylamine coated Au electrode achieved 0.07 μg/L limit of detection in river, spring and tap water samples [236]. ASV and Linear sweep voltammetry (LSV) using a mixed self-assembled monolayer of glutathione (GT), dithiothreitol (DTT) and N-acetyl-L-cysteine (NaLc) produced a good linear response in the 3-100 μg/L range with a 0.5 μg/L limit of detection [237]. Recently, Lalmalswami et al. developed an inexpensive and low cost silane grafted nanocomposite based sensor for electrochemical detection of As(III) [238]. Although, in their ASV and CV studies they achieved, an exceptional detection limit of 3.6 ng/L with spiked river water, elaborate studies with real water samples will be more encouraging.

Square wave ASV (SWASV) with Fe_3O_4 based room temperature ionic liquid (RTIL) composite electrode has shown an exceptional0.8 ng/L limit of detection for arsenic analysis in the range of 1-10 μg/L [239]. Although, CSV studies with vibrating Au microwire has shown very good sensitivity and a large linear workable range, there are significant interferences from iodide ion [240]. CSV using Hg electrode needs long accumulation times for accurate and reproducible results [241]. Cyclic voltammetry

(CV) studies with modified Au electrodes have shown very good sensitivity for arsenic determination in real water samples [242-243]. However, these CV methods with modified Au electrodes were used mostly for As(III) and also suffer from interferences from a lot of metal ions. Overall, electrochemical methods present many sensitive and cost-effective methods for arsenic analysis. Nevertheless, electrodes require a lot of attention along with regular checking and field analysis with electrodes has not proved to be robust method [244].

Table 4. Highly sensitive electrochemical techniques for arsenic analysis with field deployable abilities.

Method	Sample type	Arsenic Species	Limit of detection [μg/L].	Practical range [μg/L].	Ref.
ASV on modified Hg film	Tap/lake/ natural water	As(III)	0.3	0.5-448	[229]
ASV on modified Au electrode	Lake water	As(III)	0.02	0.2-300	[230]
ASV on a Mn coated Au microwire	Seawater	As(V)	0.0015	-	[231]
DPSAV on vibrating Au microwire	River water	As(V)	0.07	0.07-3.0	[232]
DPASV, square wave ASV on Au surface	Drinking water	As(III)	0.06	1-15	[233]
ASV on GCE with Au/Pd NPs	Laboratory solutions	As(III)	0.25	1-25	[234]
ASV on GCE (Pt/ Fe(III) MWCNT	Laboratory solutions	As(III)	0.75	7.5-22.5	[235]
ASLSV with Au-NPs	Tap/spring/ river	As(III)	0.07	1.5-225	[236]
ASV/LSV (GT+DTT+ NaLc on Au electrode)	Sea/river/lake /tap water	As(III)	0.5	3-100	[237]
CV, ASV (Silane grafted nanocomposite electrode)	Laboratory solutions	As(III)	0.0036	0.5 - 20	[238]
SWASV (Fe$_3$O$_4$-RTIL composite electrode)	Groundwater	As(III)	0.0008	1-10	[239]

CSV on vibrating Au microwire	Ground/river/ lake/tap water	As(III)	0.035	0.07-7500	[240]
CSV on Hg electrode	River/rain/ lake/tap water	As(III)	0.2	0.75-750	[241]
CV, modified Au electrode	River/tap/ ground water	As(III)	0.018	0.1-1800	[242]
CV, modified Au electrode	River/lake water	As(III)	0.047	0.1-120	[243]

4.1.7 Inductively coupled plasma mass spectrometry (ICP-MS)

ICP-MS is another standard detection technique widely used for arsenic and other metal contaminants in trace analysis [200,218]. This method can be coupled with a wide range of pretreatment and separation techniques such as hydride generation (HG), solid phase micro extraction (SPME), capillary micro extraction (CME), solvent bar micro extraction (SBME), high performance liquid chromatography (HPLC), ion chromatography (IC) and diffusive gradient in thin films (DGT). Recently, liquid-liquid micro extraction (LLME) and capillary electrophoresis (CE) are being also explored for arsenic extraction and speciation [200,218].

HPLC/IC enabled separation coupled with ICP-MS detection is one of the most common and widely used technique for arsenic testing and speciation [200]. When combined with column chromatography, HG-ICP-MS is capable of speciation of all major arsenic species in water as well as extracts obtained from solid substances. In SPME, a suitable sorbent material is coated on a solid metal or silica fiber/rod to ensure separation [245-250]. Due to availability of several sorbents combined with very less use of organic solvents, SPME-ICP-MS offer outstanding detection options. Similarly, in CME, an active sorbent layer inside a capillary serves as the extraction medium [251-252]. SBME ensures that the solvent used for extraction is restricted within a hollow fiber membrane, which is placed within the sample solution with continuous stirring. Sinking the extraction set-up within the sample solution further aids the extraction process. In case of SBME, As(V) must be pre-reduced to As(III) for better results. SBME offers better enrichment factor and faster extraction compared to CME [253]. DGT offers an in situ technique by integrating diffusive and binding gels for sampling and preconcentration using various environmental matrices. DGT methods have shown linear response over a considerable period of time and also have been used for field analysis [254-256].

ICP-MS has been widely used for arsenic analysis of water samples as well as particulate matter [200,218,257]. The minimum limit of detection achieved for water and particulate samples are 0.7ng/L and 0.05 ng/m^3 respectively [251,257]. In the estimation of arsenic (m/z = 75) using the ICP-MS technique, there can be interferences from $^{40}Ar^{35}Cl^+, ^{59}Co^{16}O$, $Sm^{2+}$and Nd^{2+} [200]. In this regard, digestion of solid particulates with chlorine containing substances such as $HClO_4$ must be avoided. Also, ^{35}Cl containing calibrating agents are discouraged, as they may combine with the inert carrier gas, Ar to produce $^{40}Ar^{35}Cl^+$ [200]. Moreover, except HG-ICP-MS, other techniques when coupled with ICP-MS are prone to matrix interferences due to variations in ionization. Laser Ablation (LA) techniques have also been used for plasma generation (LA-ICP-MS) from the sample [258]. Nevertheless, instability of the used lasers and inhomogeneity of the filters result in less reproducible results. High cost per test, maintenance cost, bulky instrumentation and requirement of skilled manpower make it less likeable for large scale testing.

4.1.8 Inductively coupled plasma- optical emission spectroscopy (ICP-OES)

ICP-OES is similar to ICP-MS, except for the fact that the emission from analyte ion is measured instead of detection of the charged ion. The optical emission spectroscopy (OES) is often referred as atomic emission spectroscopy (AES). Again, if the emission is in the form of fluorescence, then it is known as atomic fluorescence spectroscopy (AFS). Although, the ICP-OES method can be applied for water, solid and gaseous samples, most of the studies have used it for arsenic determination in atmospheric particulate matter [257]. The gaseous particulate matter is digested and converted into liquid samples. The minimum limit of detection of arsenic achieved using this method is relatively high (50 µg/L) [259]. Hence, it is suitable only for samples having high arsenic concentration. However, when coupled with HG, the limit of detection of arsenic with ICP-OES may improve to 1 µg/L [260].

4.1.9 Neutron activation analysis (NAA)

In case of NAA, the sample is subjected to neutron bombardment resulting in formation of radioactive nuclides, which undergo β and/or γ with particular half-lives and energy emissions. The measurement of the half-life and energy emitted from the arsenic or any other contaminant nuclide present in the sample provides an estimate for their concentration in water, herbal and soil samples [261-263]. It can provide accurate results over a wide range from 1 µg/L to several mg/L. This method is not only expensive but considering its radioactive nature, also requires following of a number of safety protocols by well-trained personnel.

4.1.10 Biosensors

A few bacteria-based assays have been explored for estimation of arsenic in water [264]. For instance, a spore forming bacteria namely, Bacillus subtilis was utilized for preconcentration of As(III) as well as As(V) in water under certain pH conditions. Several enzymes-based biosensors for arsenic estimation are also reported [265-266]. When acid phosphatase and polyphenol oxidase are trapped into clays and layered double hydroxides, they allow detection of As(V). The hydrolysis of phenyl phosphate into phenol by acid phosphatase, followed by the oxidation of phenol into o-quinone by polyphenol oxidase involves exchange of electrons [265]. As(V) inhibits this activity of acid phosphatase and the same inhibition rate can be monitored as function of As(V) concentration. Arsenite oxidase, is an enzyme which oxidizes arsenite to arsenate and is used for estimation of As(III) in the form of a coating on a suitable electrode [266]. The redox reactions occurring on the bio-electrode can reflect the As(III) concentration in the water. Recently, a molybdenum sulphide (MoS_2) transducer used in a liquid-ion gated field-effect transistor (FET) based biosensor was used to develop a very sensitive method for arsenic estimation which can prove very useful for potable drinking water. This method was observed to detect arsenic as low as 75 pg/L [267]. However, in this study the maximum concentration of arsenic to be estimated accurately was 0.75 µg/L, which is too low considering the amount of arsenic present in the water of the affected areas. Remarkably, copper, which generally interferes during arsenic estimation by electrochemical methods, did not interfere in the case of such biosensors. These biosensors offer a very cost effective and environment-friendly technique for arsenic estimation. Nevertheless, the performance of these biosensors over a wide range of arsenic concentrations, especially in the high concentration region is yet to be explored.

Conclusion

The choice of arsenic removal and its effectiveness depends upon the source, volume, chemical composition, pH, presence of interfering ions and the amount of arsenic in the contaminated water. Hence, prior to the implementation of any technique for arsenic removal from a particular site, it must be thoroughly tested.

Conventional arsenic removal technologies of precipitation/coprecipitation which involve addition of other chemicals are generally used for treatment of industrial effluents. However, careful pH monitoring, slow settling of precipitate and use of large flocculation tanks increase the economic burden. Several sorption techniques which use iron, iron-manganese and/or aluminium based compounds along with PRBs have proven to be very efficient and offer inexpensive solutions. Some of these most effective sorption-based techniques are listed in Table 1. However, they suffer from interferences from phosphates, silicates, carbonates, sulfates and dissolved organic materials, especially at higher pH. Disposal of the used sorbent materials also present an additional problem. At the same time, ion exchange due to their regeneration ability, effectiveness and low-cost are good treatment methods for the removal of arsenic species from water. Membrane technologies involving microfiltration, ultrafiltration, nanofiltration and reverse osmosis have proved highly effective in arsenic removal. Reverse osmosis and nanofiltration have found wide use in rural areas, with nanofiltration proving to be a better option due to its ability to work under low pressures. Magnetic treatment based arsenic removal has very little use for practical purposes. Bioremediation of arsenic offers a cost-effective green technology. However, subsequent removal of bacteria and organic wastes from the water is necessary prior to human consumption. Moreover, such methods need more research to be implemented on a large scale. Overall, ultrafiltration and nanofiltration based membrane technologies offer quick, effective, inexpensive, and easily implementable solution for arsenic removal.

Post treatment, it is ubiquitous to test the level of arsenic in the discharged water. Again, prior to human consumption, testing of arsenic in groundwater and other water sources is essential. Inexpensive colorimetric methods based on Gutzeit method present viable solution for arsenic analysis on a large scale, especially in the most affected developing countries like Nepal, Bangladesh. Most of these methods are overwhelmed by arsine leakage, operator judgement error, color degradation often result in inaccurate and irreproducible results. Other spectroscopic techniques such as LIBS, SERS, TXRF, ATR-FTIR have proved insensitive below mg/L level of arsenic. Electrochemical methods of arsenic analysis using different kinds of electrodes are observed to be extremely sensitive with limit of detection in the sub µg/L level. Although such electrode-based methods have often proved time-saving and cost effective in field analysis, the robustness of the electrodes needs to be improved. For timesaving, sensitive and accurate arsenic analysis, sophisticated and expensive techniques such as ICP-MS and HG-AAS are highly recommended. Such instruments have multi element capabilities, require skilled manpower and thus, use of such instruments for single element analysis will be unwise. Moreover, ICP-MS and HG-AAS instruments being expensive and bulky prove less useful in large scale field analysis. To conclude, electrochemical and GPCL based automated instruments along with emerging biosensor technology offer inexpensive solutions for accurate, reproducible, portable and rapid field testing of arsenic on a large scale.

References

[1] A. Shore, A. Fritsch, M. Heim, A. Schuh, M. Thoennessen, Discovery of the arsenic isotopes, At. Data. Nucl. Data Tables 96 (2010) pp. 299-306. https://doi.org/10.1016/j.adt.2009.11.001

[2] D.K. Gupta, S. Tiwari, B.H.N. Razafindrabe, S. Chatterjee, Arsenic contamination from historical aspects to the present, in D.K. Gupta, S. Chatterjee (Eds.) Arsenic contamination in the environment: The issues and solutions, Springer International Publishing AG, Switzerland, 2017, pp. 1-12. https://doi.org/10.1007/978-3-319-54356-7_1

[3] K. Henke, Arsenic: environmental chemistry, health threats and waste treatment, John Wiley & Sons Ltd, United Kingdom, 2009.

[4] D. Park, H. Yang, J. Jeong, K. Ha, S. Choi, C. Kim, C. Yoon, D. Park, D. Paek, A comprehensive review of arsenic levels in the semiconductor manufacturing industry, Ann. Occup. Hyg. 54 (2010) pp. 869-879.

[5] S. Ham, C. Yoon, S. Kim, J. Park, O. Kwon, J. Heo, D. Park, S. Choi, S. Kim, K. Ha, W. Kim, Arsenic exposure during preventive maintenance of an ion implanter in a semiconductor manufacturing factory, Aerosol Air Qual. Res. 17 (2017) pp. 990-999. https://doi.org/10.4209/aaqr.2016.07.0310

[6] International Agency for Research on Cancer, Arsenic, metals, fibres and dusts, 100C, IARC monographs on the evaluation of carcinogenic risks to humans, IARC Press, International Agency for Research on Cancer, Lyon, 2012.

[7] J.O. Nriagu, P. Bhattacharya, A.B. Mukherjee, J. Bundschuh, R. Zevenhoven, R.H. Loeppert, Arsenic in soil and groundwater: an overview, trace metals and other contaminants in the environment, Elsevier, 9 (2007) pp. 3-60. https://doi.org/10.1016/S1875-1121(06)09001-8

[8] H. Yamauchi, G. Sun, Arsenic contamination in Asia: biological effects and preventive measures, Springer Nature Singapore Pte Ltd. Singapore, 2019. https://doi.org/10.1007/978-981-13-2565-6

[9] F. Minichilli, F. Bianchi, A.M. Ronchi, F. Gorini, E. Bustaffa, Urinary arsenic in human samples from areas characterized by natural or anthropogenic pollution in Italy, Int. J. Environ. Res. Public Health, 15 (2018) pp. 299-317. https://doi.org/10.3390/ijerph15020299

[10] World Health Organization (WHO) Guidelines for drinking-water quality, 2nd edn, World Health Organization, Geneva, 1998.

[11] M. Bissen, F.H. Frimmel, Arsenic - a review. Part II: oxidation of arsenic and its removal in water treatment. Acta Hydrochim. Hydrobio. 31 (2003) pp. 97-107. https://doi.org/10.1002/aheh.200300485

[12] M.G. García, J. D'Hiriart, J. Giullitti, H. Lin, G. Custo, M.D.V. Hidalgo, M.I. Litter, M.A. Belsa, Solar light induced removal of arsenic from contaminated groundwater: the interplay of solar energy and chemical variables. Sol. Energy 77 (2004) pp. 601-613. https://doi.org/10.1016/j.solener.2004.06.022

[13] K.J. Bisceglia, K.J. Rader, R.F. Carbonaro, K.J. Farley, J.D. Mahony, D.M. Di Toro, Iron(II)-catalyzed oxidation of arsenic(III) in a sediment column. Environ. Sci. Technol. 39 (2005) pp. 9217-9222. https://doi.org/10.1021/es051271i

[14] M. J. Kim, J. Nriagu, Oxidation of arsenite in groundwater using ozone and oxygen, Sci. Total Environ. 247 (2000) pp. 71-79. https://doi.org/10.1016/S0048-9697(99)00470-2

[15] M.C. Dodd, N.D. Vu, A. Ammann, V.C. Le, R. Kissner, H.V. Pham, T.H. Cao, M. Berg, U.V. Gunten, Kinetics and mechanistic aspects of As(III) oxidation by aqueous chlorine, chloramines, and ozone: relevance to drinking water treatment, Environ. Sci. Technol. 40 (2006) pp. 3285-3292. https://doi.org/10.1021/es0524999

[16] S.J. Hug, L. Canonica, M. Wegelin, Solar oxidation and removal of arsenic at circumneutral pH in iron containing waters, Environ. Sci. Technol. 35 (2001) pp. 2114-2121. https://doi.org/10.1021/es001551s

[17] S.J. Hug, O. Leupin, Iron-catalyzed oxidation of arsenic(III) by oxygen and by hydrogen peroxide: pH-dependent formation of oxidants in the Fenton reaction, Environ. Sci. Technol. 37 (2003) pp. 2734-42. https://doi.org/10.1021/es026208x

[18] O.X. Leupin, S.J. Hug, Oxidation and removal of arsenic (III) from aerated groundwater by filtration through sand and zero-valent iron, Water Res. 39 (2005) pp. 1729-1740. https://doi.org/10.1016/j.watres.2005.02.012

[19] P.K. Dutta, S.O. Pehkonen, V.K. Sharma, A.K. Ray, Photocatalytic oxidation of arsenic(III): evidence of hydroxyl radicals, Environ. Sci. Technol. 39 (2005) pp. 1827-1834. https://doi.org/10.1021/es0489238

[20] J. Buschmann, S. Canonica, U. Lindauer, S.J. Hug, L. Sigg, Photoirradiation of dissolved humic acid induces arsenic(III) oxidation, Environ. Sci. Technol. 39 (2005) pp. 9541-9546. https://doi.org/10.1021/es051597r

[21] A. Maldonado-Reyes, C. Montero-Ocampo, O. Solorza-Feria, Remediation of drinking water contaminated with arsenic by the electro-removal process using

different metal electrodes, J. Environ. Monitor. 9 (2007) pp. 1241-1247. https://doi.org/10.1039/b708671g

[22] M. Arienzo, P. Adamo, J. Chiarenzelli, M.R. Bianco, A. De Martino, Retention of arsenic on hydrous ferric oxides generated by electrochemical peroxidation, Chemosphere 48 (2002) pp. 1009-1018. https://doi.org/10.1016/S0045-6535(02)00199-6

[23] S. Licht, X. Yu, Electrochemical alkaline Fe(VI) water purification and remediation, Environ. Sci. Technol. 39 (2005) pp. 8071-8076. https://doi.org/10.1021/es051084k

[24] H.L. Ehrlich, Bacterial oxidation of As(III) compounds, in Environmental Chemistry of Arsenic (ed. W.T.Frankenberger Jr.), Marcel Dekker, New York, 2002, pp. 313-27.

[25] M. Leist, R.J. Casey, D. Caridi, The management of arsenic wastes: problems and prospects, J. Hazard. Mater. 76 (2000) pp. 125-138. https://doi.org/10.1016/S0304-3894(00)00188-6

[26] F.-S. Zhang, H. Itoh, Photocatalytic oxidation and removal of arsenite from water using slag-iron oxide-TiO2 adsorbent, Chemosphere 65 (2006) pp. 125-131. https://doi.org/10.1016/j.chemosphere.2006.02.027

[27] T. Xu, P.V. Kamat, K.E. O'Shea, Mechanistic evaluation of arsenite oxidation in TiO2 assisted photocatalysis, J. Phys. Chem. A 109 (2005) pp. 9070-9075. https://doi.org/10.1021/jp054021x

[28] H. Lee, W. Choi, Photocatalytic oxidation of arsenite in TiO2 suspension: kinetics and mechanisms, Environ. Sci. Technol. 36 (2002) pp. 3872-3878. https://doi.org/10.1021/es0158197

[29] M.A. Ferguson, M.R. Hoffmann, J.G. Hering, TiO2-photocatalyzed As(III) oxidation in aqueous suspensions: reaction kinetics and effects of adsorption, Environ. Sci. Technol. 39 (2005) pp. 1880-86. https://doi.org/10.1021/es048795n

[30] O.S. Thirunavukkarasu, T. Viraraghavan, K.S. Subramanian, O. Chaalal, M.R. Islam, Arsenic removal in drinking water - impacts and novel removal technologies, Energy Sources, 27 (2005) pp. 209-219. https://doi.org/10.1080/00908310490448271

[31] C. Tournassat, L. Charlet, D. Bosbach, A. Manceau, Arsenic(III) oxidation by birnessite and precipitation of manganese(II) arsenate. Environ. Sci. Technol. 36 (2002) pp. 493-500.

[32] Y. Tani, N. Miyata, M. Ohashi, T. Ohnuki, H. Seyama, K. Iwahori, M. Soma, Interaction of inorganic arsenic with biogenic manganese oxide produced by a Mn-

Materials Research Forum LLC
https://doi.org/10.21741/9781644901144-4

oxidizing fungus, strain KR21-2, Environ. Sci. Technol. 38 (2004) pp. 6618-6624. https://doi.org/10.1021/es049226i

[33] M.R. Jekel, Removal of arsenic in drinking water treatment, in J.O. Nriagu (ed.) Arsenic in the Environment: Part I: Cycling and Characterization, John Wiley & Sons, Inc. New York, 1994, pp. 11-32.

[34] M.V.B. Krishna, K. Chandrasekaran, D. Karunasagar, J. Arunachalam, A combined treatment approach using Fenton's reagent and zero-valent iron for the removal of arsenic from drinking water, J. Hazard. Mater. 84 (2001) pp. 229-230. https://doi.org/10.1016/S0304-3894(01)00205-9

[35] B.A. Manning, S.E. Fendorf, B. Bostick, D.L. Suarez, Arsenic(III) oxidation and arsenic(V) adsorption reactions on synthetic birnessite, Environ. Sci. Technol. 36 (2002) pp. 976-981. https://doi.org/10.1021/es0110170

[36] S.A. Mokashi, K.M. Paknikar, Arsenic (III) oxidizing Microbacterium lacticum and its use in the treatment of arsenic contaminated groundwater, Lett. Appl. Microbiol. 34 (2002) pp. 258-262, Erratum: 35, pp. 171. https://doi.org/10.1046/j.1472-765x.2002.01083.x

[37] G.L. Anderson, P.J. Ellis, P. Khun, R. Hille, Oxidation of arsenite by Alcaligenes faecalis, W.T. Frankenberger Jr. (Ed.) in Environmental Chemistry of Arsenic, Marcel Dekker, New York, 2002, pp. 343-61.

[38] A. Majumder, M. Chaudhuri, Solar photocatalytic oxidation and removal of arsenic from ground water, Indian J. Eng. Mater. S. 12 (2005) pp. 122-128.

[39] J. Floch, M. Hideg, Application of ZW-1000 membranes for arsenic removal from water sources, Desalination 162 (2004) pp. 75-83. https://doi.org/10.1016/S0011-9164(04)00029-3

[40] D.A. Clifford, G.L. Ghurye, Metal-oxide adsorption, ion exchange, and coagulation-microfiltration for arsenic removal from water, in W.T. Frankenberger Jr (Ed.) Environmental Chemistry of Arsenic, Marcel Dekker, New York, 2002, pp. 217-245.

[41] Z. Cheng, A. van Geen, R. Louis, N. Nikolaidis, R. Bailey, Removal of methylated arsenic in groundwater with iron filings, Environ. Sci. Technol. 39 (2005) pp. 7662-7666. https://doi.org/10.1021/es050429w

[42] H.-L. Lien, R.T. Wilkin, High-level arsenite removal from groundwater by zero-valent iron, Chemosphere 59 (2005) pp. 377-386. https://doi.org/10.1016/j.chemosphere.2004.10.055

Materials Research Forum LLC
https://doi.org/10.21741/9781644901144-4

[43] N.P. Nikolaidis, G. M. Dobbs, J.A. Lackovic, Arsenic removal by zero-valent iron: field, laboratory and modeling studies, Water Res. 37 (2003) pp.1417-1425. https://doi.org/10.1016/S0043-1354(02)00483-9

[44] C. Su, R.W. Puls, Arsenate and arsenite removal by zerovalent iron: kinetics, redox transformation, and implications for in situ groundwater remediation, Environ. Sci. Technol. 35 (2001) pp. 1487-1492. https://doi.org/10.1021/es001607i

[45] S. Bang, M.D. Johnson, G.P. Korfiatis, X. Meng, Chemical reactions between arsenic and zero-valent iron in water, Water Res. 39 (2005) pp. 763-770. https://doi.org/10.1016/j.watres.2004.12.022

[46] B.A. Manning, M.L. Hunt, C. Amrhein, J.A. Yarmoff, Arsenic(III) and arsenic(V) reactions with zerovalent iron corrosion products, Environ. Sci. Technol. 36 (2002) pp. 5455-5461. https://doi.org/10.1021/es0206846

[47] S.R. Kanel, B. Manning, L. Charlet, H. Choi, Removal of arsenic(III) from groundwater by nanoscale zero-valent iron, Environ. Sci. Technol. 39 (2005) 1291-1298. https://doi.org/10.1021/es048991u

[48] G. Jegadeesan, K. Mondal, S.B. Lalvani, Arsenate remediation using nanosized modified zerovalent iron particles, Environ. Prog. 24 (2005) pp. 289-296. https://doi.org/10.1002/ep.10072

[49] O.X. Leupin, S. Hug, A.B.M. Badruzzaman, Arsenic removal from Bangladesh tube well water with filter columns containing zerovalent iron filings and sand, Environ. Sci. Technol. 39 (2005) pp. 8032-8037. https://doi.org/10.1021/es050205d

[50] K.G. Stollenwerk, J.A. Colman, Natural remediation potential of arsenic-contaminated ground water, in A.H. Welch, K.G. Stollenwerk (Eds.) Arsenic in Ground Water, Kluwer Academic Publishers, Boston, 2003, pp. 351-379. https://doi.org/10.1007/0-306-47956-7_13

[51] L. Cumbal, A.K. Sengupta, Arsenic removal using polymer-supported hydrated iron(III) oxide nanoparticles: role of Donnan membrane effect, Environ. Sci. Technol. 39 (2005) pp. 6508-6515. https://doi.org/10.1021/es050175e

[52] S.A. Ndur, D.J. Norman, Sorption of arsenic onto laterite: a new technology for filtering rural water, Geol. Soc. Am. Abst. with Programs 35 (2003) pp. 413.

[53] K.G. Stollenwerk, Geochemical processes controlling transport of arsenic in groundwater: a review of adsorption, in A.H. Welch, K.G. Stollenwerk (Eds.) Arsenic in Ground Water, Kluwer Academic Publishers, Boston, 2003, pp. 67-100. https://doi.org/10.1007/0-306-47956-7_3

[54] B.J. Lafferty, R.H. Loeppert, Methyl arsenic adsorption and desorption behavior on iron oxides, Environ. Sci. Technol. 39 (2005) pp. 2120-2127. https://doi.org/10.1021/es048701+

[55] W. Zhang, P. Singh, E. Paling, S. Delides, Arsenic removal from contaminated water by natural iron ores, Miner. Eng. 17 (2004) pp. 517-524. https://doi.org/10.1016/j.mineng.2003.11.020

[56] M. Mohapatra, S.K. Sahoo, S. Anand, R.P. Das, Removal of As(V) by Cu(II)-, Ni(II)-, or Co(II)-doped goethite samples, J. Colloid Interf. S. 298 (2006) pp. 6-12. https://doi.org/10.1016/j.jcis.2005.11.052

[57] K.A. Matis, A.I. Zouboulis, D. Zamboulis, A.V. Valtadorou, Sorption of As(V) by goethite particles and study of their flocculation, Water Air Soil Pollut. 111 (1999) pp. 297-316. https://doi.org/10.1023/A:1005088728949

[58] Y. Jia, G.P. Demopoulos, Adsorption of arsenate onto ferrihydrite from aqueous solution: influence of media (sulfate vs nitrate) added gypsum, and pH alteration, Environ. Sci. Technol. 39 (2005) pp. 9523-9527. https://doi.org/10.1021/es051432i

[59] J.A. Wilkie, J.G. Hering, Adsorption of arsenic onto hydrous ferric oxide: effects of adsorbate/adsorbent ratios and co-occurring solutes, Colloids Surf. A Physicochem. Eng. Asp. 107 (1996) pp. 97-110. https://doi.org/10.1016/0927-7757(95)03368-8

[60] X. Guo, F. Chen, Removal of arsenic by bead cellulose loaded with iron oxyhydroxide from groundwater, Environ. Sci. Technol. 39 (2005) pp. 6808-6818.

[61] X. Guo, Y. Du, F. Chen, H.-S. Park, Y. Xie, Mechanism of removal of arsenic by bead cellulose loaded with iron oxyhydroxide (β -FeOOH): EXAFS study, J. Colloid Interf. S. 314 (2007) pp. 427-433. https://doi.org/10.1016/j.jcis.2007.05.071

[62] L. Zeng, A method for preparing silica-containing iron(III) oxide adsorbents or arsenic removal, Water Res. 37 (2003) pp. 4351-4358. https://doi.org/10.1016/S0043-1354(03)00402-0

[63] J.E. Greenleaf, J.-C. Lin, A.K. Sengupta, Two novel applications of ion exchange fibers: arsenic removal and chemical-free softening of hard water, Environ. Prog. 25 (2006) pp. 300-311. https://doi.org/10.1002/ep.10163

[64] A. Joshi, M. Chaudhuri, Removal of arsenic from ground water by iron oxide-coated sand, J. Environ. Eng. 122 (1996) pp. 769-771. https://doi.org/10.1061/(ASCE)0733-9372(1996)122:8(769)

[65] I. Ko, A.P. Davis, J.-Y. Kim, K.-W. Kim, Arsenic removal by a colloidal iron oxide coated sand, J. Environ. Eng.133 (2007) pp. 891-898. https://doi.org/10.1061/(ASCE)0733-9372(2007)133:9(891)

[66] H. Genç-Fuhrman, J.C. Tjell, D. McConchie, Adsorption of arsenic from water using activated neutralized red mud, Environ. Sci. Technol. 38 (2004), 2428-2434. https://doi.org/10.1021/es035207h

[67] P. Sylvester, P. Westerhoff, T. Möller, M. Badruzzaman, O. Byod, A hybrid sorbent utilizing nanoparticles of hydrous iron oxide for arsenic removal from drinking water, Environ. Eng. Sci. 24 (2007) pp. 104-112. https://doi.org/10.1089/ees.2007.24.104

[68] S. Kundu, S.S. Kavalakatt, A. Pal, S.K. Ghosh, M. Mandal, T. Pal, Removal of arsenic using hardened paste of Portland cement: batch adsorption and column study, Water Res. 38 (2004) pp. 3780-3790. https://doi.org/10.1016/j.watres.2004.06.018

[69] O.M. Vatutsina, V.S. Soldatov, V.I. Sokolova, J. Johann, M. Bissen, A. Weissenbacher, A new hybrid (polymer/inorganic) fibrous sorbent for arsenic removal from drinking water, React. Funct. Polym. 67 (2007) pp. 184-201. https://doi.org/10.1016/j.reactfunctpolym.2006.10.009

[70] Z. Gu, and B. Deng, Use of iron-containing mesoporous carbon (IMC) for arsenic removal from drinking water, Environ. Eng. Sci. 24 (2007) pp. 113-121. https://doi.org/10.1089/ees.2007.24.113

[71] Z. Gu, B. Deng, J. Yang, Synthesis and evaluation of iron-containing ordered mesoporous carbon (FeOMC) for arsenic adsorption, Microporous Mesoporous Mater. 102 (2007) pp. 265-273. https://doi.org/10.1016/j.micromeso.2007.01.011

[72] J.A. Mũnoz, A. Gonzalo, M. Valiente, Arsenic adsorption by Fe(III)-loaded open-celled cellulose sponge, thermodynamic and selectivity aspects, Environ. Sci. Technol. 36 (2002) pp. 3405-3411.

[73] V.K. Gupta, V.K. Saini, N. Jain, Adsorption of As(III) from aqueous solutions by iron oxide-coated sand, J. Colloid Interf. S. 288 (2005) pp. 55-60. https://doi.org/10.1016/j.jcis.2005.02.054

[74] T. Yuan, J.Y. Hu, S.L. Ong, Q.F. Luo, W.J. Ng, Arsenic removal from household drinking water by adsorption, J. Environ. Sci. Heal. A 37 (2002) pp. 1721-1736. https://doi.org/10.1081/ESE-120015432

[75] L. Dupont, G. Jolly, M. Aplincourt, M. Arsenic adsorption on lignocellulosic substrate loaded with ferric ion, Environ. Chem. Lett. 5 (2007) pp. 125-29.

[76] A.L. Foster, Spectroscopic investigation of arsenic species in solid phases, in A.H. Welch, K.G. Stollenwerk (Eds.) Arsenic in Ground Water, Kluwer Academic Publishers, Boston, 2003, pp. 27-65. https://doi.org/10.1007/0-306-47956-7_2

[77] S. Chakravarty, V. Dureja, G. Bhattacharyya, S. Bhattacharjee, Removal of arsenic from groundwater using low cost ferruginous manganese ore, Water Res. 36 (2002) pp. 625-632. https://doi.org/10.1016/S0043-1354(01)00234-2

[78] E. Deschamps, V.S.T. Ciminelli, W.H. Höll, Removal of As(III) and As(V) from water using a natural Fe and Mn enriched sample, Water Res. 39 (2005) pp. 5212-5220. https://doi.org/10.1016/j.watres.2005.10.007

[79] S. Bajpai, M. Chaudhuri, Removal of arsenic from ground water by manganese dioxide-coated sand, J. Environ. Eng. 125 (1999) pp. 782-784. https://doi.org/10.1061/(ASCE)0733-9372(1999)125:8(782)

[80] I.A. Katsoyiannis, A.I. Zouboulis, M. Jekel, Kinetics of bacterial As(III) oxidation and subsequent As(V) removal by sorption onto biogenic manganese oxides during groundwater treatment, Ind. Eng. Chem. Res. 43 (2004) pp. 486-93. https://doi.org/10.1021/ie030525a

[81] J. Hlavay, K. Polýak, Determination of surface properties of iron hydroxide-coated alumina adsorbent prepared for removal of arsenic from drinking water, J. Colloid Interf. S. 284 (2005) pp. 71-77. https://doi.org/10.1016/j.jcis.2004.10.032

[82] D. Mohan, C.U. Pittman, Arsenic removal from water/wastewater using adsorbents-A critical review, J. Hazard. Mater. 142 (2007) pp. 1-53. https://doi.org/10.1016/j.jhazmat.2007.01.006

[83] Y. Kim, C. Kim, I. Choi, S. Rengaraj, J. Yi, Arsenic removal using mesoporous alumina prepared via a templating method, Environ. Sci. Technol. 38 (2004) pp. 924-931. https://doi.org/10.1021/es0346431

[84] L. Dambies, Existing and prospective sorption technologies for the removal of arsenic in water, Sep. Sci. Technol. 39 (2004) pp. 603-627. https://doi.org/10.1081/SS-120027997

[85] S. Kunzru, M. Chaudhuri, Manganese amended activated alumina for adsorption/oxidation of arsenic, J. Environ. Eng. 131 (2005) pp. 1350-1353. https://doi.org/10.1061/(ASCE)0733-9372(2005)131:9(1350)

[86] Y. Masue, R.H. Loeppert, T. A. Kramer, Arsenate and arsenite adsorption and desorption behavior on coprecipitated aluminum:iron hydroxides, Environ. Sci. Technol. 41 (2007) pp. 837-842. https://doi.org/10.1021/es061160z

[87] A.K. Dhiman, M. Chaudhuri, Iron and manganese amended activated alumina-a medium for adsorption/oxidation of arsenic from water. J. Water Supply Res. T. 56 (2007) pp. 69-74. https://doi.org/10.2166/aqua.2007.061

[88] S. Ayoob, A.K. Gupta, P.B. Bhakat, Analysis of breakthrough developments and modeling of fixed bed adsorption system for As(V) removal from water by modified calcined bauxite (MCB), Sep. Purif. Technol. 52 (2007) pp. 430-438. https://doi.org/10.1016/j.seppur.2006.05.021

[89] [G.P. Gillman, A simple technology for arsenic removal from drinking water using hydrotalcite, Sci. Total Environ. 366 (2006) pp. 926-931. https://doi.org/10.1016/j.scitotenv.2006.01.036

[90] L. Yang, Z. Shahrivari, P.K.T. Liu, M. Sahimi, T.T. Tsotsis, Removal of trace levels of arsenic and selenium from aqueous solutions by calcined and uncalcined layered double hydroxides (LDH), Ind. Eng. Chem. Res. 44 (2005) pp. 6804-6815. https://doi.org/10.1021/ie049060u

[91] S. Bang, M. Patel, L. Lippincott, X. Meng, Removal of arsenic from groundwater by granular titanium dioxide adsorbent, Chemosphere 60 (2005) pp. 389-397. https://doi.org/10.1016/j.chemosphere.2004.12.008

[92] H. Jézéquel, K. Chu, Removal of arsenate from aqueous solution by adsorption onto titanium dioxide nanoparticles, J. Environ. Sci. Heal. A 41 (2006) 1519-1528. https://doi.org/10.1080/10934520600754201

[93] C. Jing, X. Meng, S. Liu, S. Baidas, R. Patraju, C. Christodoulatos, G. P. Korfiatis, Surface complexation of organic arsenic on nanocrystalline titanium oxide, J. Colloid Interf. S. 290 (2005) pp. 14-21. https://doi.org/10.1016/j.jcis.2005.04.019

[94] B. Manna, M. Dasgupta, U.C. Ghosh, Crystalline hydrous titanium (IV) oxide (CHTO): an arsenic (III) scavenger from natural water, J. Water Supply Res. T. 53 (2004) pp. 483-495. https://doi.org/10.2166/aqua.2004.0038

[95] T. Xu, Y. Cai, K.E. O'Shea, Adsorption and photocatalyzed oxidation of methylated arsenic species in TiO2 suspensions, Environ. Sci. Technol. 41 (2007) pp. 5471-5477. https://doi.org/10.1021/es0628349

[96] P. Mondal, C.B. Majumder, B. Mohanty, Removal of trivalent arsenic (As(III)) from contaminated water by calcium chloride (CaCl2)-impregnated rice husk carbon, Ind. Eng. Chem. Res. 46 (2007) pp. 2550-2557. https://doi.org/10.1021/ie060702i

[97] B. Daus, R. Wennrich, H. Weiss, Sorption materials for arsenic removal from water: a comparative study, Water Res. 38 (2004) pp. 2948-2954. https://doi.org/10.1016/j.watres.2004.04.003

[98] M.-C. Shih, An overview of arsenic removal by pressure-driven membrane processes. Desalination 172 (2005) pp. 85-97. https://doi.org/10.1016/j.desal.2004.07.031

[99] L.V. Rajaković, The sorption of arsenic onto activated carbon impregnated with metallic silver and copper, Sep. Sci. Technol. 27 (1992) pp. 1423-1433. https://doi.org/10.1080/01496399208019434

[100] Q.L. Zhang, N.Y. Gao, Y.C. Lin, B. Xu, L.-S. Le, Removal of arsenic (V) from aqueous solutions using iron-oxide-coated modified activated carbon, Water Environ. Res. 79 (2007) pp. 931-936. https://doi.org/10.2175/106143007X156727

[101] G.T. Schmidt, N. Vlasova, D. Zuzaan, M. Kersten, B. Daus, Adsorption mechanism of arsenate by zirconyl-functionalized activated carbon, J. Colloid Interf. S. 317 (2008) pp. 228-234. https://doi.org/10.1016/j.jcis.2007.09.012

[102] Z. Gu, J. Fang, B. Deng, Preparation and evaluation of GAC-based iron-containing adsorbents for arsenic removal, Environ. Sci.Technol. 39 (2005) pp. 3833-3843. https://doi.org/10.1021/es048179r

[103] L. Yang, S. Wu, J.P. Chen, Modification of activated carbon by polyaniline for enhanced adsorption of aqueous arsenate, Ind. Eng. Chem. Res. 46 (2007) pp. 2133-2140. https://doi.org/10.1021/ie0611352

[104] L. Xiao, G.G. Wildgoose, A. Crossley, Removal of toxic metal-ion pollutants from water by using chemically modified carbon powders, Chem. Asian J. 1 (2006) pp. 614-622. https://doi.org/10.1002/asia.200600136

[105] I. Cano-Aguilera, N. Haque, G.M. Morrison, A.F. Aguilera-Alvarado, M. Gutiérrez, J.L. Gardea-Torresdey, G. de la Rosa, Use of hydride generation-atomic absorption spectrometry to determine the effects of hard ions, iron salts and humic substances on arsenic sorption to sorghum biomass, Microchem. J. 81 (2005) pp. 57-60. https://doi.org/10.1016/j.microc.2005.01.014

[106] M.N. Amin, S. Kaneco, T. Kitagawa, A. Begum, H. Katsumata, T. Suzuki, K. Ohta, Removal of arsenic in aqueous solutions by adsorption onto waste rice husk, Ind. Eng. Chem. Res. 45 (2006) pp. 8105-8110. https://doi.org/10.1021/ie060344j

[107] M.C. Teixeira, V.S.T. Ciminelli, Development of a biosorbent for arsenite: structural modeling based on X-ray spectroscopy, Environ. Sci. Technol. 39 (2005) pp. 895-900. https://doi.org/10.1021/es049513m

[108] R. Say, S. Emir, B. Garipcan, S. Patir, A. Denizli, Novel methacryloylamidophenylalanine functionalized porous chelating beads for adsorption of heavy metal ions, Adv. Polym. Tech. 22 (2003) pp. 355-364. https://doi.org/10.1002/adv.10062

[109] T. Balaji, H. Matsunaga, Adsorption characteristics of As(III) and As(V) with titanium dioxide loaded amberlite XAD-7 resin, Anal. Sci. 18 (2002) pp. 1345-1349. https://doi.org/10.2116/analsci.18.1345

[110] H. Zeng, B. Fisher, D.E. Giammar, Individual and competitive adsorption of arsenate and phosphate to a high-surface-area iron oxide-based sorbent, Environ. Sci. Technol. 42 (2008) pp. 147-152. https://doi.org/10.1021/es071553d

[111] N. Seko, F. Basuki, M. Tamada, F. Yoshii, Rapid removal of arsenic(V) by zirconium(IV) loaded phosphoric chelate adsorbent synthesized by radiation induced graft polymerization, React. Funct. Polym. 59 (2004) pp. 235-241. https://doi.org/10.1016/j.reactfunctpolym.2004.02.003

[112] N. Seko, M. Tamada, F. Yoshii, Current status of adsorbent for metal ions with radiation grafting and crosslinking techniques, Nucl. Instrum. Meth. B 236 (2005) pp. 21-29. https://doi.org/10.1016/j.nimb.2005.03.244

[113] V. Lenoble, C. Chabroullet, R. Al Shukry, B. Serpaud, V. Deluchat, J.-C. Bollinger, Dynamic arsenic removal on a MnO2-loaded resin, J. Colloid Interf. S. 280 (2004) pp. 62-67. https://doi.org/10.1016/j.jcis.2004.07.034

[114] M. Jang, E.W. Shin, J.K. Park, S.I. Choi, Mechanisms of arsenate adsorption by highly-ordered nano-structured silicate media impregnated with metal oxides, Environ. Sci. Technol. 37 (2003) pp. 5062-5070. https://doi.org/10.1021/es0343712

[115] E. Korngold, N. Belayev, L. Aronov, Removal of arsenic from drinking water by anion exchangers, Desalination, 141 (2001) pp. 81-84. https://doi.org/10.1016/S0011-9164(01)00391-5

[116] T.M. Suzuki, M.L. Tanco, D.A.P. Tanaka, Adsorption characteristics and removal of oxo-anions of arsenic and selenium on the porous polymers loaded with monoclinic hydrous zirconium oxide, Sep. Sci. Technol. 36 (2001) pp. 103-111. https://doi.org/10.1081/SS-100000854

[117] B. Manna, U.C. Ghosh, Adsorption of arsenic from aqueous solution on synthetic hydrous stannic oxide, J. Hazard. Mater. 144 (2007), pp. 522-531. https://doi.org/10.1016/j.jhazmat.2006.10.066

[118] J.V. Bothe Jr. P.W. Brown, Arsenic immobilization by calcium arsenate formation, Environ. Sci. Technol. 33 (1999) pp. 3806-3811. https://doi.org/10.1021/es980998m

[119] R.G. Robins, K. Tozawa, Arsenic removal from gold processing waste waters: the potential ineffectiveness of lime, CIM Bull. 75 (1982) pp. 171-174.

[120] K.N. Ghimire, K. Inoue, H. Yamaguchi, K. Makino, T. Miyajima, Adsorptive separation of arsenate and arsenite anions from aqueous medium by using orange waste, Water Res. 37 (2003) pp. 4945-4953. https://doi.org/10.1016/j.watres.2003.08.029

[121] C. Jing, G.P. Korfiatis, X. Meng, Immobilization mechanisms of arsenate in iron hydroxide sludge stabilized with cement, Environ. Sci. Technol. 37 (2003) pp. 5050-5056. https://doi.org/10.1021/es021027g

[122] L C. Roberts, S.J. Hug, T. Ruettimann, M.M. Billah, A.W. Khan, M.T. Rahman, Arsenic removal with iron(II) and iron(III) in waters with high silicate and phosphate concentrations, Environ. Sci. Technol. 38 (2004) pp. 307-315. https://doi.org/10.1021/es0343205

[123] T. Yuan, Q.-F. Luo, J.-Y. Hu, S.-L. Ong, W.-J. Ng, A study on arsenic removal from household drinking water, J. Environ. Sci. Heal. A 38 (2003) pp. 1731-1744. https://doi.org/10.1081/ESE-120022875

[124] J.G. Hering, P.-Y. Chen, J.A. Wilkie, M. Elimelech, Arsenic removal from drinking water during coagulation, J. Environ. Eng. 123 (1997) pp. 800-807. https://doi.org/10.1061/(ASCE)0733-9372(1997)123:8(800)

[125] B. Han, J. Zimbron, T.R. Runnells, New arsenic standard spurs search for cost-effective removal techniques, J. Am. Water Works Ass. 95 (2003) pp. 109-118. https://doi.org/10.1002/j.1551-8833.2003.tb10478.x

[126] H. Guo, D. Stüben, Z. Berner, Adsorption of arsenic(III) and arsenic(V) from groundwater using natural siderite as the adsorbent, J. Colloid Interface Sci. 315 (2007) pp. 47-53. https://doi.org/10.1016/j.jcis.2007.06.035

[127] R. Liu, X. Li, S. Xia, Y. Yangling, W. Rongchen, L. Guibai, Calcium-enhanced ferric hydroxide co-precipitation of arsenic in the presence of silicate, Water Environ. Res. 79 (2007) pp. 260-264. https://doi.org/10.2175/106143007X199324

[128] F. Sagitova, D. Bejan, N.J. Bunce, R. Miziolek, Development of an electrochemical device for removal of arsenic from drinking water, Can J. Chem. Eng. 83 (2005) pp. 889-895. https://doi.org/10.1002/cjce.5450830511

[129] D.W. Blowes, C.J. Ptacek, S. G. Benner, Treatment of inorganic contaminants using permeable reactive barriers, J. Contam. Hydrol. 45 (2000) pp. 123-137. https://doi.org/10.1016/S0169-7722(00)00122-4

[130] G. Ghurye, D. Clifford, A. Tripp, Iron coagulation and direct microfiltration to remove arsenic from groundwater, J. Am. Water Works Ass. 96 (2004) pp. 143-152. https://doi.org/10.1002/j.1551-8833.2004.tb10605.x

[131] E. Ergican, H. Gecol, A. Fuchs, The effect of co-occurring inorganic solutes on the removal of arsenic (V) from water using cationic surfactant micelles and an ultrafiltration membrane, Desalination 181 (2005) pp. 9-26. https://doi.org/10.1016/j.desal.2005.02.011

[132] Y.-H. Weng, H.C.-H. Lin, H.-H. Lee, K.C. Li, C.P. Huang, Removal of arsenic and humic substances (HSs) by electro-ultrafiltration (EUF), 122 (2005) pp. 171-176. https://doi.org/10.1016/j.jhazmat.2005.04.001

[133] H. Gecol, E. Ergican, A. Fuchs, Molecular level separation of arsenic (V) from water using cationic surfactant micelles and ultrafiltration membrane. J. Membrane Sci. 241(2004) pp. 105-119. https://doi.org/10.1016/j.memsci.2004.04.026

[134] K. Kŏsutic, L. Furăc, L. Sipos, B. Kunst, Removal of arsenic and pesticides from drinking water by nanofiltration membranes, Sep. Purif. Technol. 42 (2005) pp. 137-144. https://doi.org/10.1016/j.seppur.2004.07.003

[135] S. Xia, B. Dong, Q. Zhang, B. Xu, N. Gao, C. Causseranda, Study of arsenic removal by nanofiltration and its application in China, Desalination 204 (2007) pp. 374-379. https://doi.org/10.1016/j.desal.2006.04.035

[136] P. Brandhuber, G. Amy, Alternative methods for membrane filtration of arsenic from drinking water. Desalination 117 (1998) pp. 1-10.

[137] H. Saitúa, M. Campderrós, S. Cerutti, S. A.P. Padilla, Effect of operating conditions in removal of arsenic from water by nanofiltration membrane, Desalination, 172 (2005), pp. 173-180. https://doi.org/10.1016/j.desal.2004.08.027

[138] J.T. Mayo, C. Yavuz, S. Yean, L. Cong, H. Shiple, W. Yu, J. Falkner, A. Kan, M. Tomson, V.L. Colvin, The effect of nanocrystalline magnetite size on arsenic removal, Sci. Technol. Adv. Mat. 8 (2007) pp. 71-75. https://doi.org/10.1016/j.stam.2006.10.005

[139] Y. Zhang, M. Yang, X.-M. Dou, H. He, D.-S. Wang, Arsenate adsorption on an Fe-Ce bimetal oxide adsorbent: role of surface properties, Environ. Sci. Technol. 39 (2005) pp. 7246-7253. https://doi.org/10.1021/es050775d

[140] S. F. Lim, J. P. Chen, Synthesis of an innovative calcium-alginate magnetic sorbent for removal of multiple contaminants, Appl. Surf. Sci. 253 (2007) pp. 5772-5775. https://doi.org/10.1016/j.apsusc.2006.12.049

[141] A. Nakahira, H. Nagata, M. Takimura, K. Fukunishi, Synthesis and evaluation of magnetic active charcoals for removal of environmental endocrine disrupter and heavy metal ion, J. Appl. Phys. 101 (2007) pp. 09J114. https://doi.org/10.1063/1.2713430

[142] A. Nakahira, T. Kubo, H. Murase, Synthesis of LDH-type clay substituted with Fe and Ni ion for arsenic removal and its application to magnetic separation, IEEE T. Magn. 43 (2007) pp. 2442-2444. https://doi.org/10.1109/TMAG.2007.894359

[143] H. Okada, Y. Kudo, H. Nakazawa, A. Chiba, K. Mitsuhashi, T. Ohara, H. Wada, Removal system of arsenic from geothermal water by high gradient magnetic separation-HGMS reciprocal filter, IEEE Trans. Appl. Supercond. 14 (2004) pp. 1576-1579. https://doi.org/10.1109/TASC.2004.830718

[144] Y. Hu, J.-H. Li, Y.-G. Zhu, Y.-Z. Huang, H.-Q. Hu, P. Christie, Sequestration of As by iron plaque on the roots of three rice (Oryza sativa L.) cultivars in a low-P soil with or without P fertilizer, Environ. Geochem. Health 27 (2005) pp. 169-176. https://doi.org/10.1007/s10653-005-0132-5

[145] S.W. Al Rmalli, C.F. Harrington, M. Ayub, P.I. Haris, A biomaterial based approach for arsenic removal from water, J. Environ. Monitoring, 7 (2005) pp. 279-282. https://doi.org/10.1039/b500932d

[146] [22]. M.H. Rahman, N.M. Wasiuddin, M.R. Islam, Experimental and numerical modeling studies of arsenic removal with wood ash from aqueous streams, Can. J. Chem. Eng. 82 (2004) pp. 968-977. https://doi.org/10.1002/cjce.5450820512

[147] [23]. T.S. Anirudhan, M.R. Unnithan, Arsenic(V) removal from aqueous solutions using an anion exchanger derived from coconut coir pith and its recovery, Chemosphere, 66 (2007) pp 60-66. https://doi.org/10.1016/j.chemosphere.2006.05.031

[148] S. Mukherjee, S. Kumar, Adsorptive uptake of arsenic (V) from water by aquatic fern Salvinia natans, J. Water Supply Res. T. 54 (2005) pp. 47-53. https://doi.org/10.2166/aqua.2005.0005

[149] A. Vlyssides, E.M. Barampouti, S. Mai, Heavy metal removal from water resources using the aquatic plant Apium nodiflorum. Commun. Soil Sci. Plan. 36 (2005) pp. 1075-1081. https://doi.org/10.1081/CSS-200050499

[150] I.J. Pickering, L. Gumaelius, H.H. Harris, Localizing the biochemical transformations of arsenate in hyperaccumulating fern, Environ. Sci. Technol. 40 (2006) pp. 5010-5014. https://doi.org/10.1021/es052559a

[151] W.J. Fitz, W.W. Wenzel, H. Zhang, J. Nurmi, K. Štipek, Z. Fischerova, P. Schweiger, G. Köllensperger, L.Q. Ma, G. Stingeder, Rhizosphere characteristics of the arsenic hyperaccumulator Pteris vittata L. and monitoring of phytoremoval efficiency, Environ. Sci. Technol. 37 (2003) pp. 5008-5014. https://doi.org/10.1021/es0300214

[152] L.L. Embrick, K.M. Porter, A. Pendergrass, D.J. Butcher, Characterization of lead and arsenic contamination at Barber Orchard, Haywood County, NC, Microchem. J. 81 (2005) pp. 117-121. https://doi.org/10.1016/j.microc.2005.01.007

[153] P.R. Baldwin, D.J. Butcher, Phytoremediation of arsenic by two hyperaccumulators in a hydroponic environment, Microchem. J. 85 (2007) pp. 297-300. https://doi.org/10.1016/j.microc.2006.07.005

[154] C.-Y. Wei, C. Wang, X. Sun, W.-Y. Wang, Arsenic accumulation by ferns: a field survey in southern China, Environ. Geochem. Health, 29 (2007) pp. 169-177. https://doi.org/10.1007/s10653-006-9046-0

[155] J.W. Huang, C.Y. Poynton, L.V. Kochian, M.P. Elless, Phytofiltration of arsenic from drinking water using arsenic-hyperaccumulating ferns, Environ. Sci. Technol. 38 (2004), pp. 3412-3417. https://doi.org/10.1021/es0351645

[156] J. Pratas, M.N.V. Prasad, H. Freitas, L. Conde, Plants growing in abandoned mines of Portugal are useful for biogeochemical exploration of arsenic, antimony, tungsten and mine reclamation, J. Geochem. Explor. 85 (2005) pp. 99-107.

[157] M. Montes-Bayón, J. Meija, D.L. LeDuc, N. Terry, J.A. Caruso, A. Sanz-Medel, HPLC-ICP-MS and ESI-Q-TOF analysis of biomolecules induced in Brassica juncea during arsenic accumulation, J. Anal. At. Spectrom. 19 (2004) pp. 153-158. https://doi.org/10.1039/B308986J

[158] T. Budinova, N. Petrov, M. Razvigorova, J. Parra, P. Galiatsatou, Removal of As(III) from aqueous solution by activated carbons prepared from solvent extracted olive pulp and olive stones, Ind. Eng. Chem. Res. 45 (2006) pp. 1896-1901. https://doi.org/10.1021/ie051217a

[159] C.-C. Chen, Y.-C. Chung, Arsenic removal using a biopolymer chitosan sorbent, J. Environ. Sci. Heal. A 41 (2006) pp. 645-658. https://doi.org/10.1080/10934520600575044

[160] R. Say, N. Yilmaz, A. Denizli, Biosorption of cadmium, lead, mercury, and arsenic ions by the fungus penicillium purpurogenum, Sep. Sci. Technol. 38 (2003) pp. 2039-2053. https://doi.org/10.1081/SS-120020133

[161] I. A. Katsoyiannis, A. I. Zouboulis, Application of biological processes for the removal of arsenic from groundwaters, Water Res. 38 (2004) pp. 17-26. https://doi.org/10.1016/j.watres.2003.09.011

[162] I. Katsoyiannis, A. Zouboulis, H. Althoff, H. Bartel, As(III) removal from groundwaters using fixed-bed upflow bioreactors, Chemosphere 47 (2002) pp. 325-332. https://doi.org/10.1016/S0045-6535(01)00306-X

[163] J. Kostal, R. Yang, C.H. Wu, A.Mulchandani, W. Chen, Enhanced arsenic accumulation in engineered bacterial cells expressing ArsR, Appl. Environ. Microbiol. 70 (2004) pp. 4582-4587.

[164] T. Ohnuki, F. Sakamoto, N. Kozai, T. Ozaki, T. Yoshida, I. Narumi, E. Wakai, T. Sakai, A.J. Francis, Mechanisms of arsenic immobilization in a biomat from mine discharge water, Chem. Geol. 212 (2004) pp. 279-290.

[165] M.X. Loukidou, K.A. Matis, A.I. Zouboulis, M. Liakopoulou-Kyriakidou, Removal of As(V) from wastewaters by chemically modified fungal biomass, Water Res. 37 (2003) pp. 4544-4552. https://doi.org/10.1016/S0043-1354(03)00415-9

[166] H.J. Reisinger, D.R. Burris, J.G. Hering, Remediating subsurface arsenic contamination with monitored natural attenuation, Environ. Sci. Technol. 39 (2005) pp. 458A-464A. https://doi.org/10.1021/es053388c

[167] Information available on www.sensafe.com (accessed 20.05.20).

[168] Information available on www.wagtech.co.uk (accessed 01.04.14).

[169] Information available on http://www.vitasalus.net/purtest-arsenic-water-test-kit-1-test-kit (accessed 20.05.20).

[170] Information available on http://www.hach.com/arsenic-low-range-test-kit/product?id=7640217303 (accessed 20.05.20).

[171] Information available on http://www.merckmillipore.com/food-analytics/rapid-arsenic-tests/c_Hzib.s1OprIAAAEbFfcXP9oy (accessed 20.05.20).

[172] H. Gutzeit, Pharmaz. Zeit. 24 (1879) pp. 263.

[173] H. Gutzeit, The quantitative determination of Arsenic, Pharmaz. Zeit. 36 (1891) pp. 748-756.

[174] T. Akter, F.T. Jhohura, F. Akter, T.R. Chowdhury, S.K. Mistry, D. Dey, M.K. Barua, M.A. Islam, M. Rahman, Water Quality Index for measuring drinking water quality in rural Bangladesh: a cross-sectional study, J. Health Popul. Nutr. 35 (2016) pp. 4(1-12). https://doi.org/10.1186/s41043-016-0041-5

[175] M.M.H. Khan, K. Aklimunnessa, M. Kabir, M. Mori, Determinants of drinking arsenic-contaminated tubewell water in Bangladesh, Health Policy Plann. 22 (2007) pp. 335-343. https://doi.org/10.1093/heapol/czm018

[176] B. Das, M.M. Rahman, B. Nayak, A. Pal, U.K. Chowdhury, S.C. Mukherjee, K.C. Saha, S. Pati, Q. Quamruzzaman, D. Chakraborti, Groundwater arsenic contamination, its health effects and approach for mitigation in West Bengal, India and Bangladesh, Water Qual. Expos. Hea. 1 (2009) pp. 5-21. https://doi.org/10.1007/s12403-008-0002-3

[177] A.M.R. Chowdhury, M. Jakariya, Science, 284 (1999) pp. 1621 https://doi.org/10.1126/science.284.5420.1621e

[178] M.M. Rahman, D. Mukherjee, M.K. Sengupta, U.K. Chowdhury, D. Lodh, C.R. Chanda, S. Roy, M. Selim, Q. Quamruzzaman, A.H. Milton, S.M. Shahidullah, M.T. Rahman, D. Chakraborti, Effectiveness and reliability of arsenic field testing kits: are the million dollar screening projects effective or not?, Environ. Sci. Technol. 36 (2002) pp. 5385-5394. https://doi.org/10.1021/es020591o

[179] R.R. Reddy, G.D. Rodriguez, T.M. Webster, M.J. Abedin, M.R. Karim, L. Raskin, K.F. Hayes, Evaluation of arsenic field test kits for drinking water: recommendations for improvement and implications for arsenic affected regions such as Bangladesh, Water Res. 170 (2020) pp. 115325 (1-9). https://doi.org/10.1016/j.watres.2019.115325

[180] M. Salman, M. Athar, W.U. Zaman, U. Shafique, J. Anwar, R. Rehman, S. Ameer, M. Azeem, Micro-determination of arsenic in aqueous samples by image scanning and computational quantification, Anal. Methods 4 (2012) pp. 242-246. https://doi.org/10.1039/c1ay05569k

[181] E. Truog, A.H. Meyer, Improvements in the Deniges colorimetric method for phosphorus and arsenic, Ind. Eng. Chem. Anal. Ed. 1 (1929) pp. 136-139. https://doi.org/10.1021/ac50067a011

[182] P.K. Dasgupta, H. Huang. G. Zhang G. P. Cobb, Photometric measurement of trace As(III) and As(V) in drinking water, Talanta, 58 (2002) pp.153-64.

[183] S. Hu, J. Lu, C. Jing, A novel colorimetric method for field arsenic speciation analysis, J. Environ. Sci. 24 (2012) pp. 1341-1346. https://doi.org/10.1016/S1001-0742(11)60922-4

[184] K. Toda, T. Ohba, M. Takaki, S. Karthikeyan, S. Hirata, P.K. Dasgupta, Speciation-capable field instrument for the measurement of arsenite and arsenate in water, Anal. Chem. 77 (2005) pp. 4765-4773. https://doi.org/10.1021/ac050193e

[185] H. Bode, K. Hachmann, Photometric arsenic determination with silver diethyldithiocarbamate, Z. Anal. Chem. 229 (1967) pp. 261-266. https://doi.org/10.1007/BF00512979

[186] H. Bode, K. Hachmann, The reaction between arsine and silver diethyldithiocarbamate, Z. Anal. Chem. 241 (1968) pp. 18-30. https://doi.org/10.1007/BF00527733

[187] B.W. Budesinsky, Arsenic colorimetry with silver diethyldithiocarbamate, Microchem. J. 24 (1979) pp. 80- 87. https://doi.org/10.1016/0026-265X(79)90041-9

[188] K. Morita, E. Kaneko, Spectrophotometric determination of arsenic in water samples based on micro particle formation of ethyl violet-molybdoarsenate, Anal. Sci. 22 (2006) pp. 1085-1089. https://doi.org/10.2116/analsci.22.1085

[189] J.R. Kalluri, T. Arbneshi, S.A. Khan, A. Neely, P. Candice, B. Varisli, M. Washington, S. McAfee, B. Robinson, S. Banerjee, A.K. Singh, D. Senapati, P.C. Ray, Use of gold nanoparticles in a simple colorimetric and ultrasensitive dynamic light scattering assay: selective detection of arsenic in groundwater, Angew. Chem. 48 (2009) pp. 9668-9671. https://doi.org/10.1002/anie.200903958

[190] Y. Wu, L. Liu, S. Zhan, F. Wang, P. Zhou, Ultrasensitive aptamer biosensor for arsenic(III) detection in aqueous solution based on surfactant-induced aggregation of gold nanoparticles, Analyst 137 (2012) pp. 4171-4178. https://doi.org/10.1039/c2an35711a

[191] Z.-X. Lin, L. Chang, J. Li, L.-M. Liu, Determination of As in industrial wastewater by laser-induced breakdown spectroscopy, Spectrosc. Spect. Anal. 29 (2009) pp. 1675-1677.

[192] Z. Xu, J. Hao, F. Li, X. Meng, Surface-enhanced Raman spectroscopy of arsenate and arsenite using Ag nanofilm prepared by modified mirror reaction, J. Colloid Interf. S. 347 (2010) pp. 90-95. https://doi.org/10.1016/j.jcis.2010.03.028

[193] M. Mulvihill, A. Tao, K. Benjauthrit, J. Arnold, P. Yang, Surface-enhanced Raman spectroscopy for trace arsenic detection in contaminated water, Angew. Chem. Int. Ed. 47 (2008) pp. 6456-6460. https://doi.org/10.1002/anie.200800776

[194] A. Prange, A. Knöchel, W. Michelis, Multi-element determination of dissolved heavy metal traces in sea water by total-reflection X-ray fluorescence spectrometry, Anal. Chim. Acta. 172 (1985) pp. 79-100. https://doi.org/10.1016/S0003-2670(00)82596-9

[195] B. Staniszewski, P. Freimann, A solid phase extraction procedure for the simultaneous determination of total inorganic arsenic and trace metals in seawater: sample preparation for total-reflection X-ray fluorescence, Spectrochim. Acta B 63 (2008) pp. 1333-1337. https://doi.org/10.1016/j.sab.2008.08.018

[196] R. Juvonen, A. Parviainen, K. Loukola-Ruskeeniemi, Evaluation of a total reflection X-ray fluorescence spectrometer in the determination of arsenic and trace metals in environmental samples, Geochem. Explor. Env. A. 9 (2009) pp. 173-178. https://doi.org/10.1144/1467-7873/09-205

[197] H. Barros, L.-M. M. Parra, L. Bennun, E.D. Greaves, Determination of arsenic in water samples by total reflection X-ray fluorescence using pre-concentration with alumina, Spectrochim. Acta B 65 (2010) pp. 489-492. https://doi.org/10.1016/j.sab.2010.04.004

[198] B. McAuley, S. E. Cabaniss, Quantitative detection of aqueous arsenic and other oxoanions using attenuated total reflectance infrared spectroscopy utilizing iron oxide coated internal reflection elements to enhance the limits of detection, Anal. Chim. Acta, 581 (2007) pp. 309-317. https://doi.org/10.1016/j.aca.2006.08.023

[199] E.S. Forzani, K. Foley, P. Westerhoff, N. Tao, Detection of arsenic in groundwater using a surface plasmon resonance sensor, Sens. Actuators B Chem. 123 (2007) pp. 82-88. https://doi.org/10.1016/j.snb.2006.07.033

[200] J. Ma, M.K. Sengupta, D. Yuan, P.K. Dasgupta, Speciation and detection of arsenic in aqueous samples: A review of recent progress in non-atomic spectrometric methods, Anal. Chim. Acta, 831 (2014) pp.1-23. https://doi.org/10.1016/j.aca.2014.04.029

[201] W. Wang, Y. Lv, X. Hou, A potential visual fluorescence probe for ultratrace arsenic (III) detection by using glutathione-capped CdTe quantum dots, Talanta 84 (2011) pp. 382-386. https://doi.org/10.1016/j.talanta.2011.01.012

[202] N. Butwong, T. Noipa, R. Burakham, S. Srijaranai, W. Ngeontae, Determination of arsenic based on quenching of CdS quantum dots fluorescence using the gas-diffusion

flow injection method, Talanta 85 (2011) pp. 1063-1069.
https://doi.org/10.1016/j.talanta.2011.05.023

[203] M.S. Hosseini, S. Nazemi, Preconcentration determination of arsenic species by sorption of As(V) on Amberlite IRA-410 coupled with fluorescence quenching of L-cysteine capped CdS nanoparticles, Analyst 138 (2013) pp. 5769-5776.
https://doi.org/10.1039/c3an00869j

[204] A.U. Rehman, M. Yaqoob, A. Waseem, A. Nabi, Determination of arsenic (V) in freshwaters by flow injection with luminol chemiluminescence detection, Int. J. Environ. Anal. Chem. 88 (2008) pp. 603-612.
https://doi.org/10.1080/03067310801912103

[205] W. Som-aum, H. Li, J. Liu, J.-M. Lin, Determination of arsenate by sorption pre-concentration on polystyrene beads packed in a microfluidic device with chemiluminescence detection, Analyst 133 (2008) pp. 1169-1175.
https://doi.org/10.1039/b801608a

[206] K. Fujiwara, Y. Watanabe, K. Fuwa, J.D. Winefordner, Gas phase chemiluminescence with ozone oxidation for the determination of arsenic, antimony, tin, and selenium, Anal. Chem. 54 (1982) pp. 125-128.
https://doi.org/10.1021/ac00238a035

[207] C. Lomonte, M. Currell, R.J.S. Morrison, I.D. McKelvie, S.D. Kolev, Sensitive and ultra-fast determination of arsenic(III) by gas-diffusion flow injection analysis with chemiluminescence detection, Anal. Chim. Acta 583 (2007) pp. 72-77.
https://doi.org/10.1016/j.aca.2006.09.049

[208] M.K. Sengupta, M.F. Sawalha, S.I. Ohira, A.D. Idowu, P.K. Dasgupta, Green analyzer for the measurement of total arsenic in drinking water: electrochemical reduction of arsenate to arsine and gas phase chemiluminescence with ozone, Anal. Chem. 82 (2010) pp. 3467-3473. https://doi.org/10.1021/ac100604y

[209] M.K. Sengupta, P.K. Dasgupta, Oxidation state-differentiated measurement of aqueous inorganic arsenic by continuous flow electrochemical arsine generation coupled to gas-phase chemiluminescence detection, Anal. Chem. 83 (2011) pp. 9378-9383. https://doi.org/10.1021/ac201972m

[210] A.D. Idowu, P.K. Dasgupta, G. Zhang, K. Toda, J.R. Garbarino, A gas-phase chemiluminescence based analyzer for waterborne arsenic, Anal. Chem. 78 (2006) pp. 7088-7097. https://doi.org/10.1021/ac061439y

[211] A.K. Ghosh, A.N. Das, P.K. Dasgupta, A fast, accurate, speciation-capable, automated, and green gas-phase chemiluminescence approach for analyzing waterborne arsenic, LCGC Special Issues, 33 (2015) pp. 10-17.

[212] M.K. Sengupta, Z.A. Hossain, S.I. Ohira, P.K. Dasgupta, A simple inexpensive gas phase chemiluminescence analyzer for measuring trace levels of arsenic in drinking water, Environ. Pollut. 158 (2010) pp. 252-257. https://doi.org/10.1016/j.envpol.2009.07.014

[213] R. García, A. P. Báez, Atomic Absorption Spectrometry (AAS) in M. A. Farrukh (Ed.) Atomic Absorption Spectroscopy, IntechOpen, 2012. Retrieved from: https://www.intechopen.com/books/atomic-absorption-spectroscopy/atomic-absorption-spectrometry-aas- https://doi.org/10.5772/25925

[214] J. Dedina, D.L. Tsalev, Hydride Generation Atomic Absorption Spectrometry, John Wiley & Sons Inc. Chichester, 1995, pp. 182-245.

[215] R.S. Braman, C.C. Foreback, Methylated forms of arsenic in the environment, Science, 182 (1973) pp. 1247-1249. https://doi.org/10.1126/science.182.4118.1247

[216] A.G. Howard, (Boro)hydride techniques in trace element speciation, J. Anal. At. Spectrom. 12 (1997) pp. 267-272. https://doi.org/10.1039/a605050f

[217] W. Goessler, D. Kuehnelt, Analytical methods for the determination of arsenic and arsenic compounds in the environment, in W.T. Franjzenberzer (Ed.), Environmental Chemistry of Arsenic, Marcel Dekker, Netherlands, 2001, pp. 27-50.

[218] H.M. Anawar, Arsenic speciation in environmental samples by hydride generation and electrothermal atomic absorption spectrometry, Talanta, 88 (2012) pp. 30-42. https://doi.org/10.1016/j.talanta.2011.11.068

[219] F. Shemirani, M. Baghdadi, M. Ramezani, Preconcentration and determination of ultra trace amounts of arsenic (III) and arsenic (V) in tap water and total arsenic in biological samples by cloud point extraction and electro thermal atomic absorption spectrometry, Talanta 65 (2005) pp. 882-887. https://doi.org/10.1016/j.talanta.2004.08.009

[220] H. Jiang, B. Hu, B. Chen, L. Xia, Hollow fiber liquid phase microextraction combined with electrothermal atomic absorption spectrometry for the speciation of arsenic (III) and arsenic (V) in fresh waters and human hair extracts, Anal. Chim. Acta 634 (2009) pp. 15-21. https://doi.org/10.1016/j.aca.2008.12.008

[221] M. Ghambarian, M. R. Khalili-Zanjani, Y. Yamini, A. Esrafili, N. Yazdanfar, Preconcentration and speciation of arsenic in water specimens by the combination of

solidification of floating drop microextraction and electrothermal atomic absorption spectrometry, Talanta, 81 (2010) pp. 197-201. https://doi.org/10.1016/j.talanta.2009.11.056

[222] R. E. Rivas, I. López-García, M. Hernández-Córdoba, Speciation of very low amounts of arsenic and antimony in waters using dispersive liquid-liquid microextraction and electrothermal atomic absorption spectrometry, Spectrochim. Acta B 64 (2009) pp. 329-333. https://doi.org/10.1016/j.sab.2009.03.007

[223] S. Gunduz, S. Akman, A. Baysal, M. Kahraman, The use of silver nanoparticles as an effective modifier for the determination of arsenic and antimony by electrothermal atomic absorption spectrometry, Spectrochim. Acta B 65 (2010) pp. 297-300. https://doi.org/10.1016/j.sab.2010.03.011

[224] S. Gunduz, S. Akman, A. Baysal, M. Culha, The use of gold nanoparticles as an effective modifier for the determination of arsenic and antimony by electrothermal atomic absorption spectrometry, Microchim. Acta 172 (2011) pp. 403-407. https://doi.org/10.1007/s00604-010-0500-4

[225] I.B. Karadjova, P.K. Petrov, I. Serafimovski, T. Stafilov, D.L. Tsalev, Arsenic in marine tissues - the challenging problems to electrothermal and hydride generation atomic absorption spectrometry, Spectrochim. Acta B 62 (2007) pp. 258-268. https://doi.org/10.1016/j.sab.2006.10.008

[226] E.C. Lima, J.L. Brasil, J.C.P. Vaghetti, Evaluation of different permanent modifiers for the determination of arsenic in environmental samples by electrothermal atomic absorption spectrometry, Talanta 60 (2003) pp. 103-113. https://doi.org/10.1016/S0039-9140(03)00046-8

[227] J.H.T. Luong, E. Majid, K.B. Male, Analytical tools for monitoring arsenic in the environment, The Open Analytical Chemistry Journal 1 (2007) pp. 7-14. https://doi.org/10.2174/1874065000701010007

[228] D.E. Mays, A. Hussam, Voltammetric methods for determination and speciation of inorganic arsenic in the environment - a review, Anal. Chim. Acta 646 (2009) pp. 6-16. https://doi.org/10.1016/j.aca.2009.05.006

[229] A.A. Ensafi, A.C. Ring, I. Fritsch, Highly sensitive voltammetric speciation and determination of inorganic arsenic in water and alloy samples using ammonium 2-amino-1-cyclopentene-1-dithiocarboxylate, Electroanalysis 22 (2010) pp. 1175-1185. https://doi.org/10.1002/elan.200900347

[230] D. Li, J. Li, X. Jia, Y. Han, E. Wang, Electrochemical determination of arsenic(III) on mercaptoethylamine modified Au electrode in neutral media, Anal. Chim. Acta 733 (2012) pp. 23-27. https://doi.org/10.1016/j.aca.2012.04.030

[231] K. Gibbon-Walsh, P. Salaün, C.M.G. van den Berg, Determination of arsenate in natural pH seawater using a manganese-coated gold microwire electrode, Anal. Chim. Acta 710 (2012) pp. 5-57. https://doi.org/10.1016/j.aca.2011.10.041

[232] G.M.S. Alves, J.M.C.S. Magalhães, P. Salaün, C.M.G. van den Berg, H.M.V.M. Soares, Simultaneous electrochemical determination of arsenic, copper, lead and mercury in unpolluted fresh waters using a vibrating gold microwire electrode, Anal. Chim. Acta 703 (2011) pp. 1-7. https://doi.org/10.1016/j.aca.2011.07.022

[233] A. Giacomino, O. Abollino, M. Lazzara, M. Malandrino, E. Mentasti, Determination of As(III) by anodic stripping voltammetry using a lateral gold electrode: experimental conditions, electron transfer and monitoring of electrode surface, Talanta 83 (2011) pp. 1428-1435. https://doi.org/10.1016/j.talanta.2010.11.033

[234] Y. Lan, H. Luo, X. Ren, Y. Wang, Y. Liu, Anodic stripping voltammetric determination of arsenic(III) using a glassy carbon electrode modified with gold-palladium bimetallic nanoparticles, Microchim. Acta 178 (2012) pp. 153-161. https://doi.org/10.1007/s00604-012-0827-0

[235] S.-H. Shin, H.-G. Hong, Anodic stripping voltammetric detection of arsenic(III) at platinum-iron(III) nanoparticle modified carbon nanotube on glassy carbon electrode, B. Korean Chem. Soc. 31 (2010) pp. 3077-3083. https://doi.org/10.5012/bkcs.2010.31.11.3077

[236] T. Gu, L. Bu, Z. Huang, Y. Liu, Z. Tang, Y. Liu, S. Huang, Q. Xie, S. Yao, X. Tu, X. Luo, S. Luo, Dual-signal anodic stripping voltammetric determination of trace arsenic(III) at a glassy carbon electrode modified with internal-electrolysis deposited gold nanoparticles, Electrochem. Commun. 33 (2013) pp. 43-46. https://doi.org/10.1016/j.elecom.2013.04.019

[237] L. Chen, N. Zhou, J. Li, Z. Chen, C. Liao, J. Chen, Synergy of glutathione, dithiothreitol and N-acetyl-L-cysteine self-assembled monolayers for electrochemical assay: sensitive determination of arsenic(III) in environmental and drinking water, Analyst 136 (2011) pp. 4526-4532. https://doi.org/10.1039/c1an15454k

[238] J. Lalmalsawmi, Zirliangnura, D. Tiwari, S.-M. Lee, Low cost, highly sensitive and selective electrochemical detection of arsenic(III) using silane grafted based

nanocomposite, Environ. Eng. Res. 25 (2020) pp. 579-587. https://doi.org/10.4491/eer.2019.245

[239] C. Gao, Y.-Y. Yu, S.-Q. Xiong, J.-H. Liu, X.-J. Huang, Electrochemical detection of arsenic(III) completely free from noble metal: Fe3O4 microspheres-room temperature ionic liquid composite showing better performance than gold, Anal. Chem. 85 (2013) pp. 2673-2680. https://doi.org/10.1021/ac303143x

[240] K. Gibbon-Walsh, P. Salaün, C.M.G. van den Berg, Arsenic speciation in natural waters by cathodic stripping voltammetry, Anal. Chim. Acta 662 (2010) pp. 1-8. https://doi.org/10.1016/j.aca.2009.12.038

[241] M. Grabarczyk, Stripping voltammetric determination of As(III) in natural water samples containing surface active compounds, Electroanalysis 22 (2010) pp. 2017-2023. https://doi.org/10.1002/elan.201000056

[242] S.S. Hassan, Sirajuddin, A.R. Solangi, T.G. Kazi, M.S. Kalhoro, Y. Junejo, Z.A. Tagar, N.H. Kalwar, Nafion stabilized ibuprofen-gold nanostructures modified screen printed electrode as arsenic(III) sensor, J. Electroanal. Chem. 682 (2012) pp. 77-82. https://doi.org/10.1016/j.jelechem.2012.07.006

[243] J.-F. Huang, H.-H. Chen, Gold-nanoparticle-embedded nafion composite modified on glassy carbon electrode for highly selective detection of arsenic(III) Talanta 116 (2013) pp. 852-859. https://doi.org/10.1016/j.talanta.2013.07.063

[244] H. Huang, P.K. Dasgupta, A field-deployable instrument for the measurement and speciation of arsenic in potable water, Anal. Chim. Acta 380 (1999) pp. 27-37. https://doi.org/10.1016/S0003-2670(98)00649-7

[245] P. Mondal, C. Balomajumder, B. Mohanty, Quantitative separation of As(III) and As(V) from a synthetic water solution using ion exchange columns in the presence of Fe and Mn ions, Clean Soil Air Water 35 (2007) pp. 255-260. https://doi.org/10.1002/clen.200700002

[246] S. Chen, X. Zhan, D. Lu, C. Liu, L. Zhu, Speciation analysis of inorganic arsenic in natural water by carbon nanofibers separation and inductively coupled plasma mass spectrometry determination, Anal. Chim. Acta 634 (2009) pp. 192-196. https://doi.org/10.1016/j.aca.2008.12.018

[247] S. Chen, C. Liu, M. Yang, D. Lu, L. Zhu, Z. Wang. Solid-phase extraction of Cu, Co and Pb on oxidized single-walled carbon nanotubes and their determination by inductively coupled plasma mass spectrometry, J. Hazard Mater. 170 (2009) pp. 247-251. https://doi.org/10.1016/j.jhazmat.2009.04.104

[248] N.B. Issa, V.N. Rajakovič-Ognjanovič, B.M. Jovanovič, L.V. Rajakovič, Determination of inorganic arsenic species in natural waters - benefits of separation and preconcentration on ion exchange and hybrid resins, Anal. Chim. Acta 673 (2010) pp. 185-193. https://doi.org/10.1016/j.aca.2010.05.027

[249] N. B. Issa, V. N. Rajakovič-Ognjanovič, A. D. Marinkovič, L. V. Rajakovič, Separation and determination of arsenic species in water by selective exchange and hybrid resins, Anal. Chim. Acta 706 (2011) pp. 191-198. https://doi.org/10.1016/j.aca.2011.08.015

[250] E. Boyacı, A. Çağir, T. Shahwan, A. E. Eroğlu, Synthesis, characterization and application of a novel mercapto- and amine-bifunctionalized silica for speciation/sorption of inorganic arsenic prior to inductively coupled plasma mass spectrometric determination, Talanta 85 (2011) pp. 1517-1525. https://doi.org/10.1016/j.talanta.2011.06.021

[251] W. Hu, F. Zheng, B. Hu, Simultaneous separation and speciation of inorganic As(III)/As(V) and Cr(III)/Cr(VI) in natural waters utilizing capillary microextraction on ordered mesoporous Al2O3 prior to their on-line determination by ICP-MS, J. Hazard. Mater. 151 (2008) pp. 58-64. https://doi.org/10.1016/j.jhazmat.2007.05.044

[252] F. Zheng, B. Hu, Dual silica monolithic capillary microextraction (CME) on-line coupled with ICP-MS for sequential determination of inorganic arsenic and selenium species in natural waters, J. Anal. At. Spectrom. 24 (2009) pp. 1051-1061. https://doi.org/10.1039/b900057g

[253] X. Pu, B. Chen, B. Hu, Solvent bar microextraction combined with electrothermal vaporization inductively coupled plasma mass spectrometry for the speciation of inorganic arsenic in water samples, Spectrochim. Acta B, 64 (2009) pp. 679-684. https://doi.org/10.1016/j.sab.2009.06.001

[254] W.W. Bennett, P.R. Teasdale, J.G. Panther, D.T. Welsh, D.F. Jolley, New diffusive gradients in a thin film technique for measuring inorganic arsenic and selenium(IV) using a titanium dioxide based adsorbent, Anal. Chem. 82 (2010) pp. 7401-7407. https://doi.org/10.1021/ac101543p

[255] W.W. Bennett, P.R. Teasdale, J.G. Panther, D.T. Welsh, D.F. Jolley, Speciation of dissolved inorganic arsenic by diffusive gradients in thin films: selective binding of As(III) by 3-mercaptopropyl-functionalized silica gel, Anal. Chem. 83 (2011) pp. 8293-8299. https://doi.org/10.1021/ac202119t

[256] W.W. Bennett, P.R. Teasdale, D.T. Welsh, J.G. Panther, R.R. Stewart, H.L. Price, D.F. Jolley, Inorganic arsenic and iron(II) distributions in sediment porewaters

investigated by a combined DGT-colourimetric DET technique, Environ. Chem. 9 (2012) pp. 31-40. https://doi.org/10.1071/EN11074

[257] D. Sánchez-Rodas, A.S. de la Campa, L. Alsioufi, Analytical approaches for arsenic determination in air: a critical review, Anal. Chim. Acta 898 (2015) pp. 1-18. https://doi.org/10.1016/j.aca.2015.09.043

[258] D. Pozebon, G.L. Scheffler, V.L. Dressler, M.A.G. Nunes, Review of the applications of laser ablation inductively coupled plasma mass spectrometry (LA-ICP-MS) to the analysis of biological samples, J. Anal. At. Spectrom. 29 (2014) pp. 2204-2228. https://doi.org/10.1039/C4JA00250D

[259] F. Halek, M.K. Rad, R.M. Darbani, A. Kavousirahim, Concentrations and source assessment of some atmospheric trace elements in northwestern region of Tehran, Iran, Bull, Environ. Contam. Toxicol. 84 (2010) pp. 185-190. https://doi.org/10.1007/s00128-009-9902-6

[260] P.K. Pandey, K.S. Patel, P. Šubrt, Trace elemental composition of atmospheric particualte at Bhilai in central-east India, Sci. Total Environ. 215 (1998) pp. 123-134. https://doi.org/10.1016/S0048-9697(98)00111-9

[261] M.T. Islam, S.A. Islam, S.A. Latif, Detection of arsenic in water, herbal and soil samples by neutron activation analysis technique, Bull. Environ. Contam. Toxicol. 79 (2007) pp. 327-330. https://doi.org/10.1007/s00128-007-9209-4

[262] J. Sano, Y. Kikawada, T. Oi, Determination of As(III) and As(V) in hot spring and river waters by neutron activation analysis with pyrrolidinedithiocarbamate coprecipitation technique, J. Radioanal. Nucl. Chem. 278 (2008) pp. 111-116. https://doi.org/10.1007/s10967-007-7149-4

[263] W. M. Sanchez, B. Zwicker, A. Chatt, Determination of As(III) As(V) MMA and DMA in drinking water by solid phase extraction and neutron activation, J. Radioanal. Nucl. Chem. 282 (2009) pp. 133-138. https://doi.org/10.1007/s10967-009-0224-2

[264] E. Diesel, M. Schreiber, J.R. van der Meer, Development of bacteria-based bioassays for arsenic detection in natural waters, Anal. Bioanal. Chem. 394 (2009) pp. 687-693. https://doi.org/10.1007/s00216-009-2785-x

[265] S. Cosnier, C. Mousty, X. Cui, X. Yang, S. Dong, Specific determination of As(V) by an acid phosphatase-polyphenol oxidase biosensor, Anal. Chem. 78 (2006) pp. 4985-4989. https://doi.org/10.1021/ac060064d

Materials Research Forum LLC
https://doi.org/10.21741/9781644901144-4

[266] K.B. Male, S. Hrapovic, J.M. Santini, J.H.T. Luong, Biosensor for arsenite using arsenite oxidase and multiwalled carbon nanotube modified electrodes, Anal. Chem. 79 (2007) 7831-7837. https://doi.org/10.1021/ac070766i

[267] J.H. An, J. Jang, Highly sensitive FET-type aptasensor using flower-like MoS2 nanospheres for real-time detection of arsenic(III), Nanoscale 9 (2017) pp. 7483-7492. https://doi.org/10.1039/C7NR01661A

[268] D. Melamed, Monitoring arsenic in the environment: a review of science and technologies with the potential for field measurements, Anal. Chim. Acta, 532 (2005) pp. 1-13. https://doi.org/10.1016/j.aca.2004.10.047

[269] J. Tyson, The determination of arsenic compounds: a critical review, Int. Sch. Res. Notices 2013 (2013) pp. 835371. https://doi.org/10.1155/2013/835371

[270] M.S. Reid, K.S. Hoy, J.R.M. Schofield, J.S. Uppal, Y. Lin, X. Lu, H. Peng, X.C. Le, Arsenic speciation analysis: A review with an emphasis on chromatographic separations, Trac-Trend Anal. Chem. 123 (2020) pp. 115770. https://doi.org/10.1016/j.trac.2019.115770

[271] Information available on http://www.euro.who.int/__data/assets/pdf_file/0014/123071/AQG2ndEd_6_1_Arseni c.PDF (accessed on 22.04.2020)

Materials Research Forum LLC
https://doi.org/10.21741/9781644901144-5

Chapter 5

The Applicability of Eggshell Waste as a Sustainable Biosorbent Medium in Wastewater Treatment – A Review

P. Musonge[1,2,*], C. Harripersadth[1]

[1] Institute of Systems Science, Durban University of Technology, Durban, South Africa

[2] Faculty of Engineering, Mangosuthu University of Technology, Durban, South Africa

paulm@dut.ac.za, musonge.paul@mut.ac.za, c.harripersadth@gmail.com

Abstract

The exemplary properties of eggshell waste have gained a lot of attention due to its chemical composition and bio-degradable features making it a suitable choice to be used in wastewater treatment. The use of biosorption as an alternate treatment technology to conventional processes such as chemical precipitation and ion exchange is seen as a promising solution to the many drawbacks experienced by conventional processes. Furthermore, due to higher imposed environmental legislations, eco-friendly and low-cost considerations have set the momentum in the search for biosorbents of this nature. With the circular economy being the focal point of industrial operations, eggshell waste is a highly promising biosorbent due to its non-toxicity properties and its ability to be converted from a waste material to a valuable product. In this review paper, fundamental aspects of biosorption will be discussed where the main focus will lie in qualitatively examining the properties of eggshell waste, binding mechanisms, kinetics and isotherm modelling that make it an attractive option to be used in the biosorptive process. Finally, a summary of the important considerations for future research work in this field is presented.

Keywords

Water Treatment, Bio-Materials, Eggshell Waste, Sustainability, Circular Economy

Contents

1.   Introduction

In recent decades, the marked increase in industrial activities and water usage worldwide have resulted in the release of several contaminants into the aquatic environment, such as toxic heavy metals, dyes and pesticides, amongst others [1]. Aquatic bodies are continually being degraded through the discharge of industrial wastewaters and domestic wastes [2]. With that said, anthropogenic activities are the most common sources where hazardous contaminants are distributed into the environment through the continual discharge of sewage and industrial effluent [3]. Thus, governments have established environmental restrictions with regards to the quality of wastewater, forcing industries to treat waste effluent before being discharged [4]. Currently, there are several treatment technologies used for the removal of hazardous pollutants which include chemical precipitation, ion exchange, reverse osmosis and ultrafiltration to name a few, all of which have been reviewed by several researchers [5-7].

However, the use of these processes come with drawbacks where they are not recommended for applications with low concentrations of metal ions [8]. They become technically inefficient when the target concentration falls below 100 ppm. Additionally, some of these technologies require large quantities of reagents, have high energy requirements not to mention the production of large amounts of secondary waste sludge [1]. Thus, the adsorption process has been the focus of research studies in recent times for the removal of toxic contaminants. The main attraction of this process is its cost effectiveness and good removal performance [9].

Commercially, there are available adsorbents such as activated carbon and zeolites however it has restricted use due to its high costs [10] leaving a gap in the market. This gap has led to the investigation of materials of agricultural and biological origin [9]. Raw and natural agricultural wastes are among some of the cheap adsorbents that can be used. One such agricultural-based adsorbent is eggshell waste.

Statistically, China was the leading egg producing country worldwide in a 2016 report, producing 453.17 billion eggs (The Statistics Portal) with the United States producing 101.95 billion cases followed by India, Mexico and Brazil with 83, 54 and 46 billion tons and 20.8 million cases being produced in South Africa in 2014 [11, 12]. Due to the large quantities of egg production worldwide, a considerable quantity of shell residue is generated, which is considered as waste [13]. Landfill sites generally reject eggshell waste due to the protein component of the membrane which attracts vermin [14]. Additionally, disposal costs (mainly on landfill sites) are quite significant and are expected to rise due to the continual increase in landfill taxes [15]. Thus, several research studies have been conducted on eggshell waste to source alternative uses to remediate this worldwide concern [13, 15, 16]. Possible applications range from applications as fertilizers and animal feed to the adsorption of heavy metals, paper treatment, catalysts for biodiesel production, production of hydrolyzed protein to bone and dental implants to name a few [13]. Proposing the use of eggshell waste as a biosorbent serves a twofold solution, it lessens the impact of environment pollution while simultaneously providing an environmentally friendly treatment technology to be used in wastewater treatment.

Eggshells are an attractive option due to their low costs, abundant availability in nature, non-toxic properties, large specific surface areas, and high potential of ion exchange for charged pollutants [17]. Sorption by eggshells occurs mainly by an exchange reaction making it suitable to be used as a biological sorbent of metal ions [14]. Currently,

Advances in Wastewater Treatment I   Materials Research Forum LLC
Materials Research Foundations **91** (2021) 171-   https://doi.org/10.21741/9781644901144-5

eggshells have few industrial applications despite being a plentiful waste, and several potential applications are being investigated due to its calcium carbonate structure [18]. This review paper concentrates on the use of eggshell waste as the adsorbent medium for the process of biosorption. The chemical and physical properties of eggshells make them exemplary materials to be used in the biosorption process.

1.1   Characterization of eggshell waste

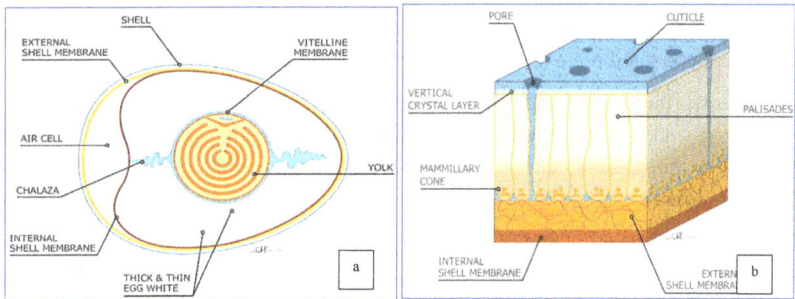

Figure 1. (a) Longitudinal section to depict the interior contents of a chicken egg and (b) Cross-sectional view, Adapted from [19].

The hen shell represents approximately 11% of the total weight (approximately 60 g) of the egg [15]. It is comprised of a calcified shell and shell membrane layer including both inner and outer membranes [20]. There are three layers, the outermost layer called the cuticle, the calcium carbonate layer called the testa, and the innermost layer called the mammillary layer. The cuticle and mammillary layers both form a matrix composed of protein fibers bonded to calcite (calcium carbonate) crystal. The two layers are also constructed such that there are numerous circular openings (pores), which allow the transpiration of water and gaseous exchanges, throughout the shell. Each eggshell has been estimated to contain between 7000 and 17,000 pores [15] making it an excellent choice to be used as an adsorbent.

The outer surface of the shell is covered with a mucin protein, which acts as a soluble plug for the pores in the shell [21] with the organic matter of the shell and shell membrane containing proteins as major constituents with small amounts of carbohydrates and lipids [20].

Advances in Wastewater Treatment I   Materials Research Forum LLC
Materials Research Foundations **91** (2021) 171-   https://doi.org/10.21741/9781644901144-5

The membrane layer of the eggshell is an amorphous natural biomaterial with an intricate lattice of stable and water insoluble fibers [22]. There are two membranes which lie directly beneath the shell called the inner and outer shell membrane. The outer membrane remains adhered to the mammillary layer of the shell, while the inner membrane surrounds the liquid of the egg. The two membranes separate at the larger end of the egg and create a space between them called the "Air Cell". The membranes are composed of protein fiber arranged to form a semi-permeable membrane with a total thickness of around 100 μm [21]. The membrane surface bears positively charged sites produced by basic side chains of amino acids. It has a very high surface area with functional groups such as hydroxyl (−OH), thiol (−SH), carboxyl (−COOH), amino (−NH2) and amide (−CONH2) to name a few [22].

The mineralized shell constitutes about 96% calcium carbonate with the remaining components comprising of the organic matrix (2%), magnesium, phosphorus and a variety of trace elements [19, 22]. The membrane constitutes nearly 60% protein of which 35 % is collagen, 10% glucosamine, 9% chondroitin and 5 % hyaluronic acid together with other inorganic components like Ca, Mg, Si, Zn, etc. in minor quantities. The density of the shell is approximately 2.53 g/cm3, which is significantly larger than that of the membrane which is 1.358 g/cm3 [22].

2.   Modifications

There are many ways in which eggshell waste can be prepared for use in the biosorptive process however several researchers prefer modifications to enhance sorption capabilities [23-27]. Having said that, using it in its natural form without any physical or chemical modifications have also proven to be effective [14, 18, 28-31].

The most commonly used modification process of eggshell waste is achieved through calcination. Calcination is a form of thermal modification which can be performed using several methodologies achieved by subjecting the eggshell biomass to a furnace at a specified temperature. This provides several advantageous characteristics to the eggshell, such as increased surface area, pore volumes and an enriched CaO content which facilitates in the ion exchange process between the surface of the biosorbent and the metal ion/dye. The temperature range generally used varies between 600-900 °C [23, 32].

The goal of calcination is to dissociate calcium carbonate into calcium oxide and carbon dioxide as explained by the following reaction:

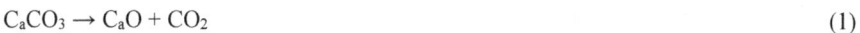

$$C_aCO_3 \rightarrow C_aO + CO_2 \hspace{5cm} (1)$$

In doing so, the high calcium oxide content achieved through calcination provides the necessary sites to bind heavy metals and dyes through the mechanism of ion exchange as explained by various researchers [32-34].

The method of Slimani et al. [35] involved calcination at 900 °C for 2 hr to produce a composition of 61.95% Ca, 0.79% Mg, 0.65% Si, 0.64% Na, 0.47% P, 0.17% Al, 0.12% Cl with a surface area of 62.42 m^2/g. In contrast to this, 800 °C for a period of 4 hr was used by Al-Ghouti et al. [33] which produced a CaO content of 58.31% (3.02%, C and 53.33% O_2), 0.09% SiO_2, 0.02% Al_2O_3, 0.01% MgO, 0.02% Fe_2O_3, 0.10% Na_2O and 0.16% P_2O_5 respectively. Interestingly, a very low surface area of 1.81 m^2/g was achieved compared to the raw form of 1.15 m^2/g using this methodology.

In the methodology of Witoon et al. [34], eggshells were calcined at 900 °C for 1 h under a N_2 atmosphere producing 97.42% CaO, 1.63 % MgO, 0.52 % P_2O_5, 0.26 % SO_3, 0.08% K_2O, 0.05 % SrO, 0.02% Cl, 0.01% Fe_2O_3 and 0.01% CuO respectively. Following this methodology produced a reasonably high surface area of 13.45 m^2/g in comparison to the raw form of 0.05 m^2/g.

Generally, a pre-screening run should be conducted to determine a suitable temperature for calcination. The raw eggshell should be dried at a low and high range temperature to determine which temperature yields the highest weight loss. Thereafter, this temperature is recommended to be used until a constant weight is achieved, at which, the constant weight temperature should be used to ensure maximum conversion to CaO.

Although the term modification generally implies alteration by thermal or chemical means, another form of modifying the properties of eggshell waste includes mechanical activation, as demonstrated by Baláž et al. [36]. Significant increase of the sorption capacity of the eggshell biomaterial toward Cd(II) was observed upon milling, as evidenced by the maximum monolayer value of 329 mg/g. Baláž et al. [37] attributed the main driving force to be the presence of the aragonite phase as a consequence of phase transformation from calcite which occurred during milling.

Modification by converting eggshell waste to CHAP which is carbonate hydroxylapatite, is also effective in achieving high biosorptive capacities. Zheng et al. [38] used eggshell waste synthesized to CHAP to investigate the removal of Cd (II) and Cu (II) from aqueous solutions. The mechanisms proposed were ion exchange and surface adsorption which produced maximum capacities of 111.1 mg/g Cd and 142.86 mg/g Cu corresponding to a 94% and 93.17 % removal efficiency respectively. Using synthesized CHAP from eggshells is highly effective however this methodology consists of costly chemicals and thermal treatments as demonstrated by Ramesh et al. [39], who produced hydroxyapatite by a sintering method. In the above process, calcined eggshell and dicalcium hydrogen phosphate di-hydrate was mixed followed by a heat treatment at 800 °C to produce a flower-like powder which was then subjected to a thermal treatment in the range of 1050 – 1350 °C for 2 hr.

The aim of biosorption is to source a cost- effective biosorbent and the use of thermal/chemical modifications defeat this purpose. A trade-off is generally needed to obtain reasonable removal capacities through the use of pre-treatment processes of moderate costs. In this respect, mechanical modification is the preferred choice to costly thermal and chemical treatment processes.

Table 1. Modifications of eggshell waste.

No	Type of Modification	Modifying Agents	Adsorbate Types	References
1	Thermal	Furnace	Basic yellow 28	[35]
2	Thermal	Furnace	Boron	[33]
3	Thermal	Furnace	Carbon dioxide	[34]
4	Thermal	Furnace	Carbon dioxide	[40]
5	Thermal	Furnace	Cyanide	[23]
6	Thermal	Furnace	Lead	[41]
7	Thermal	Furnace	Carbon dioxide	[42]
8	Thermal	Furnace	Phosphorus	[43]
9	Thermal	Furnace	Phenol	[44]
10	Thermal	Furnace	Phosphorous	[45]
11	Thermal	Furnace	Basic blue 9 Acid orange 51	[46]
12	Thermal	Furnace	Phosphate	[47]
13	Thermal	Furnace	Chromium	[48]
14	Thermal	Furnace	Cadmium Chromium Lead	[49]
15	Chemical	H_3PO_4 Calcium hydroxide	Cadmium Copper	[38]
16	Chemical	Iron oxide	Copper	[50]
17	Chemical	Methanol Hydrochloric acid	Boron	[33]

18	Chemical	FeCl$_3$·6H2O Potassium hydroxide sodium alginate calcium chloride	Congo red	[51]
19	Chemical	Hydrochloric acid (0.5%)	Cyanide	[52]
20	Chemical	Iron chloride	Phosphate	[25]
21	Chemical	Magnesium nitrate Aluminium nitrate Urea Hydrochloric acid	Chromium	[53]
22	Chemical	Titanium tetraisopropoxide Hydrochloric acid Cadmium acetate dehydrate ammonium sulphide	Methylene blue	[24]
23	Chemical	5% polyvinyl alcohol 0.5% sodium alginate Calcium chloride Boric acid	Reactive red dye	[54]
24	Chemical	40 % acetic acid Ammonia nickel chloride	Copper	[55]
25	Chemical	nitric acid phosphoric acid ammonium hydroxide	Nickel	[56]
26	Chemical	Phosphoric acid Calcium hydroxide	Lead	[26]
27	Thermal, chemical	goethite, a-MnO$_2$ and goethite/a-MnO$_2$	Arsenate	[57]
28	Thermal, chemical	Calcium chloride Sodium alginate Ammonia	Carbon dioxide	[58]

29	Thermal, chemical	Nitric acid (NH4)₃PO₄	Copper Zinc Lead	[59]
30	Thermal, chemical	Furnace,dimethylformamide (DMF)	Lead	[60]
31	Thermal, chemical	Furnace ferric sulphate	pathogenic bacteria and antibiotic resistance genes	[61]
32	Thermal, chemical	Furnace Sulphuric acid Sodium hydroxide Aluminium nitrate Aluminium sulphate.	Fluoride	[62]
33	Thermal, chemical	Furnace Nitric acid Sulphuric acid	Congo red	[63]
34	Thermal, chemical	Furnace Sodium hydroxide Hydrochloric acid Sodium citrate monohydrate, Sodium dihydrogen phosphate Sodium tetraborate Glycine Nitric acid	Lead	[64]
35	Physical	Ball mill	Cadmium	[36, 65]

3. Isotherm modelling

Biosorption equilibrium data are often required to develop an effective and accurate model that describes the behavior of pollutant sequestration from aqueous media. Isotherm modelling provides useful information relating the maximum adsorption capacity to the possible interactions between an adsorbent and adsorbate [66]. Among the various models that exist (Langmuir, Frendlich, Redlich-Peterson, Dubinin-

Radushkevich, Elovich, Sips, Extended Langmuir, Extended Sips etc.), the most commonly adopted models by researchers include the Langmuir and Freundlich models [67-70]. These models are used to identify the adsorption performance and adsorption mechanism of adsorbents or adsorbates. With respect to biosorption using eggshell biomass, the adsorption equilibrium can be satisfactorily represented by the Langmuir isotherm [28, 35, 38, 57]. In the Langmuir model, the distribution of adsorbate on a biosorbent (qe) is graphed against the equilibrium concentration (Ce) in solution [67]. The Langmuir isotherm describes a homogenous monolayer adsorption, where all the active sites (identical and in a fixed number) have an equal affinity for the sorbate, with no interaction between the sorbate molecules in the plane of the surface [71]. On the other hand, the Freundlich model is an empirical model developed for adsorption at lower concentrations on heterogeneous surfaces and assumes the adsorbed molecules interact with their adjacent neighbors [72]. The works of Al-Ghouti et al. [33] demonstrated the applicability of this model. The model equation of this model states that the adsorption energy is reduced exponentially with the decreasing in the number of active sites of an adsorbent [73].

4. Kinetic modelling

In order to investigate and determine the mechanism of biosorption and any potential rate-controlling steps such as mass transport and chemical reaction processes, several kinetic models are proposed [74]. However, there are three models that can adequately describe the biosorption process which includes the pseudo-1st order, pseudo 2nd order and intraparticle diffusion equation [75-77]. The pseudo-first order model was proposed by Lagergren in 1898 which describes the rate of sorption to be proportional to the number of sites unoccupied by the solutes [78]. With respect to the pseudo second order kinetics model, Qaiser et al. [79], states that biosorption is based on a chemical reaction, involving the exchange of electrons between the biosorbent and metal. In intraparticle diffusion, there is a migration of metal ions from the liquid phase to the solid phase [80]. In this model, the sorption capacity is related to the diffusion constant which changes with time and the boundary layer thickness is proportional to the intercept of this model. When it equates to zero, it implies that intraparticle diffusion is the rate controlling step in biosorption.

5. Mechanism of biosorption

The mechanism of biosorption regarding the affinity of a pollutant to a biosorbent is very complex and there are many ways for a pollutant to bind/accumulate to a biosorbent. The

Materials Research Forum LLC

https://doi.org/10.21741/9781644901144-5

complex structure of raw biomass implies that there are many ways, by which these biosorbents remove pollutants, but these are not yet fully understood [81].

The mechanism by which adsorption occurs may be classified according to the dependence on the cell structure as suggested by Veglio et al. [82] which can be metabolism dependent or non – metabolism dependent. In their work, they proposed that the location where the pollutant is removed determines the type of removal mechanism that occurs. For example, extracellular accumulation/precipitation, cell surface sorption/precipitation and Intracellular accumulation.

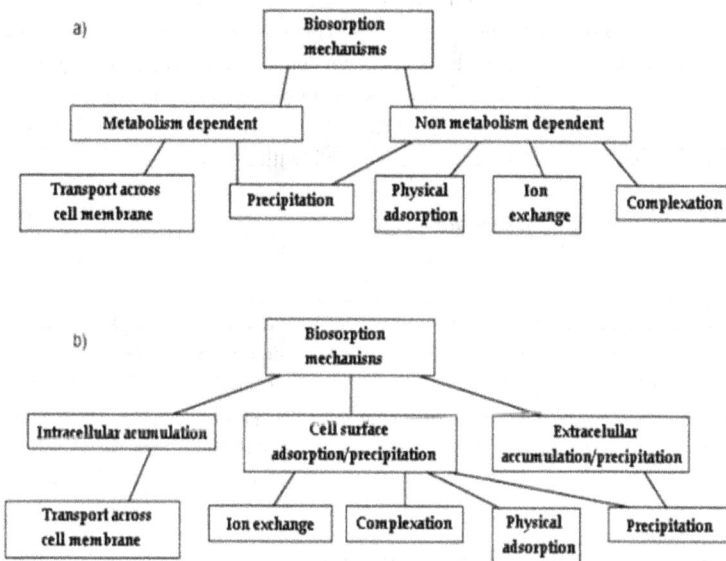

Figure 2. Biosorption mechanisms as classified by [82]. (a) – Classified according to the dependence on the cellular metabolism. (b) – Classified according to the location where biosorption occurs.

Agricultural based wastes are classified as non – metabolism dependent and so the removal mechanisms include precipitation reactions, physical adsorption, ion exchange and complexation to name a few. They normally contain a variety of organic compounds (lignin, cellulose and hemicelluloses) and functional groups such as hydroxyl, carbonyl and amino [83] capable of binding pollutants. Both organic compounds and functional groups have great affinity for metal ion complexation [84]. Eggshell waste consists of carbonates due to their calcium carbonate structure which favors metal ion binding [85].

As explained by Putra et al. [83], eggshell waste mainly consists of calcium carbonate which binds metal ions such as Cu, Pb and Zn through the electrostatic interaction between the metal and the carbonate group. Additionally, the mechanism of complexation was found to occur through electron pair sharing between the electron donor atoms (N and O). The mechanism proposed by Flores-Cano et al. [18] was precipitation and ion exchange in the removal of Cd using eggshell waste. With respect to dyes, ionic interactions between sulfonate groups in dyes and positively charged sites on the surface of eggshell seem to be the dominating mechanism of removal according to Tsai et al. [46]. Supplementing this theory, cationic dyes are explained through the electrical double layer mechanism whilst that of the anionic dyes favor the electrostatic attraction on the positively charged surface of eggshell. It can be said that the mechanism of removal in biosorption is highly dependent on the pollutant being removed and the surface chemistry of a biosorbent with environmental factors such as solution chemistry playing an important role. For example, as reported by Vijayaraghavan et al. [86], eggshell was used as an additive to precipitate Pb within a pH range of 2- 5.

6.   Factors Affecting biosorption

6.1   pH

Solution pH is a fundamental parameter that affects the speciation of metal ions in solution through hydrolysis, complexation, and redox reactions [87]. Among all other variables, pH is most critical since it not only influences the speciation of metal ions but also the charge on the sorption site of a biomass [88]. Thus, it is vital to consider the ionic state of a functional group in the biomass together with the metal solution chemistry at different pH values. With a change in pH, the behavior of a functional group will change accordingly.

Adsorption due to complexation is pH dependent which occurs at some specific pH [89] and when the pH of a solution changes, the complex formation will be affected resulting in a change in biosorption efficiency of the biomass. In low pH environments, metal ions are generally positively charged and are attracted by negatively charged biomass. As the pH of a solution is increased, the amount of hydroxide ions is increased. In general, metal ions are precipitated out in the alkaline pH range and do not contribute towards biosorption which inherently highlights the upper limit of pH to be used.

6.2   Concentration

The effect of concentration affects the sorption capacity of a biosorbent with a higher concentration resulting in a higher solute uptake. At low solute concentrations, the ratio of moles of solute to available surface area is low and the resultant fraction sorbed becomes independent of the initial metal concentration [87]. At higher concentrations, the sites available for adsorption to occur are fewer compared to the number of moles of solute present, hence, the removal of solute strongly depends on the initial solute concentration.

Many researchers propose that a lower biosorption capacity is obtained with an increase in metal ion concentration due to insufficient surface area to accommodate higher concentrations of metal ions available in solution [90]. At lower concentrations, metal ions present in solution interact with the binding sites and thus the percentage biosorption is higher than those at higher initial metal ion concentrations. At higher concentrations, lower biosorption yield is due to the saturation of biosorption sites. As a result, diluting effluent containing high metal ion concentrations can increase the purification yield [90]. Malakondaiah et al. [90] used hen egg shell to remove Cu from a concentration range of 20-100 mg/L with an increase in metal uptake observed at increased initial metal ion concentration with a decrease in biosorption capacity from 89.25% – 74.84%. It was proposed that metal uptake increased due to the increase in driving force that is, the concentration gradient.

The concentration of solute in the adsorbate is an important parameter, which determines both the utilization and feasibility of a biosorbent in a biosorption process. Jin et al. [67] suggests experiments to be carried out at the maximum possible initial solute concentration to attain the highest saturation potential of a biosorbent.

6.3   Contact time

Heavy metal biosorption is metabolism-independent and typically occurs rapidly, in particular, for the uptake of cationic metal ions [91]. As demonstrated by Senthilkumar et al. [92], a contact time of 120 minutes was sufficient to achieve equilibrium in the removal of Cu with rapid uptake observed within the first 120 min within the range of 5-360 min that was varied, achieving an 84.41 % removal capacity. In this regard, Jai and co-workers compared the efficiencies of both raw and calcined eggshell only to find that the natural form performed better for Pb removal achieving an 86% removal within the first 10 min. On the other hand, complete removal of Cd and a 99% removal of Cr was achievable with the calcined form with rapid uptake observed within the first 10 min. Generally, biosorption equilibrium of cationic metals are reached within the first 60-120 min with rapid uptake being observed at the initial stages [93-95]. With respect to dye removal, equilibrium time was found to be concentration dependent in a study conducted by Slimani et al., 2014 who investigated the removal of basic yellow 28 dye onto calcined eggshells. Equilibrium was achieved in 45 min for the ranges of concentration

40–60 mg/L and about 120 min for the ranges of concentration 20–30 mg/L. This was in agreement with the work conducted by Ehrampoush et al. [31] who investigated the removal of reactive red dye. According to Abdel-Khalek et al. [96] cationic and anionic dye removal is comprised of two phases: a primary rapid phase and a second slow phase which were corroborated by the results of Kobiraj et al. [29] and Zulfikar et al. [30].The equilibrium time of cationic and anionic dyes using whole eggshell matrix (eggshell + membrane) exhibited an initial rapid phase lasting 40 min followed by a plateau around 50 min. In elaborating on dyes, 40 minutes was sufficient to reach equilibrium in the removal of methyl orange using thermally treated eggshell [97]. The adsorption rate was rapid during the first 5 min and then continued at a slower rate from 5 to 35 min, and reached a plateau after 40 min. In contrast, 180 minutes was required for the removal of Reactive red dye in the work of Elkady et al. [54]. Overall, it can be concluded that a shorter equilibrium time is required for the elimination of dyes in comparison to heavy metals with dyes generally requiring between 5-60 min at most.

6.4   Adsorbent mass

The biosorbent mass strongly influences the extent of adsorption in a biosorption process. An increase in the biomass concentration increases the amount of solute adsorbed due to the increased number of binding sites. On the other hand, the quantity of biosorbed solute per unit weight of biosorbent decreases with increasing biosorbent dosage due to complex interactions [87]. Sometimes, at high sorbent loads, the available solute is insufficient to completely cover the available exchangeable sites on the sorbent surface resulting in slow uptake. As suggested by Barka et al. [76], a further increase in biomass concentration above a certain dosage does not lead to a significant improvement in biosorption yield explained as a consequence of partial aggregation of biomass, resulting in a decrease in effective specific surface area.

7.   Eggshell waste for heavy metal removal

In the last decade, the use of poultry waste in wastewater treatment processes has received much attention from researchers all over the world [22]. The removal of heavy metals from waste effluent is a matter of great interest due to the water scarcity and declining quality of water worldwide. Numerous metals such as lead, copper, chromium, cadmium, lead, arsenic etc. are known to be significantly toxic and in recent years their effective removal using eggshell biomass as adsorbents have been achieved [18, 23, 28, 49, 83, 92].

Operating parameters such as concentration, dosage and pH amongst others greatly affects the efficacy of a biomass and it is important to investigate these variables to find the optimal conditions which will give the highest percentage removal. In this regard, a study conducted by Putra and colleagues revealed optimal conditions at pH 6.0, 0.1 g biomass with 90 min equilibrium time in the removal of Cu, Pb and Zn ions [83]. Mezenner and Bensmaili [25], however, reported optimal conditions at a biomass dosage of 10 g/L, pH 7, 60 min contact time with an initial concentration of 2.8 mg/L for the removal of phosphate. In a study by Yeddou and Bensmaili [95], the optimum conditions were found to be 2.5 g/L biomass dosage, pH 6, 40 min contact time, 10 mg/L solution at a temperature of 20 °C for the removal of Fe(II) ions.

It is generally agreed that optimal conditions depend on the type of biomass used, adsorbate being removed as well as environmental conditions that prevail as can be demonstrated by a study conducted by Kanyal et al. [98] who varied the contact time, agitation speed and pH to investigate the removal efficiency of Cu and Pb ions. The optimized conditions concluded from this experiment was obtained at a pH of 7, 100 rev/min with a contact time of 90 minutes. In this regard, process parameters in an investigation by Mashangwa et al. [99] revealed optimal conditions being attainable at pH 7, biomass dose of 7 g with a contact time of 360 min (for the removal of 100 ppm metal ions) in the removal of Zn, Pb, Cu and Ni ions from synthetic solutions and several metals from three acid mine drainage (AMD) samples. Under these conditions, a 97% removal for Pb, 95% for Cu, 94% for Ni, and 80% for Zn was achievable signifying the applicability of eggshells as a biosorbent. It is also of note that many other metals inclusive of Al, Fe, K, Ni, and Zn, exhibited removal capacities greater than 75 % in AMD sample 1. Additionally, K had a 98.78% adsorption, while Mg, Sr, and Zn had 72.33, 68.75, and 53.07% capacities respectively in sample 2. In sample 3, As, Cr, Cu, Fe, antimony, and tellurium ions were greater than 75%.

Particle size is another variable which affects the removal capacity as highlighted by a study investigating the removal of divalent Pb cations by Vijayaraghavan et al. [86] who used eggshell as an additive to precipitation. On reducing the particle size from 750 to 100 microns, a significant increase in removal efficiency from 30.7 to 99.6% was achieved with 35 min being sufficient to achieve equilibrium.

Characterisation studies form a vital component in understanding the interaction between a biosorbent and adsorbate in an adsorption process. Putra et al. [83] used the techniques of Scanning Electron Microscope (SEM), Energy Dispersive X-ray Spectrometer (EDX) and Fourier Transform Infrared Spectrometer (FTIR) to characterise their biomass which revealed interactions with metal ions resulting in the formation of discrete aggregates on the biosorbent surface. Ion exchange proceeded through the electrostatic attraction between the metal ion and carbonate group whilst the complexation mechanism involved electron pair sharing between electron donor atoms (O and N). Results from this study proposed that carbonate, carbonyl, hydroxyl and amine groups were the main adsorption sites in eggshell. In a study involving the sorption mechanism of Cd(II), Flores-Cano et al. [18] attributed the main mechanisms to the calcareous layer of the eggshell, but also slightly on the membrane layer. It was demonstrated that the sorption process was not reversible with the main mechanisms of removal being attributed to precipitation and ion exchange. The precipitation of $(Cd,Ca)CO_3$ on the surface of the eggshell was corroborated by SEM and XRD analysis where the biosorbent was characterized by several techniques which confirmed the calcite phase of the eggshell due to its $CaCO_3$ structure. In elaborating on Cd removal, research conducted by Baláž et al. [37] revealed a significant increase in adsorption upon milling with the main driving force for adsorption proven to be the presence of the aragonite phase as a consequence of phase transformation from the calcite phase which occurred during milling. Cd was found to be adsorbed in a non-reversible way, as documented by XRD and EDX measurements [100]. Ok et al.[101] investigated the effectiveness of eggshell waste on the immobilization of Cd and Pb using techniques such as SEM and XRD to characterize the eggshell biomass.

Table 2. Biosorptive capacity of heavy metals using eggshell waste.

No	Type of Modification	Modifying Agents	Adsorbate Types	References
1	Thermal	Furnace	Basic yellow 28 dye	[35]
2	Thermal	Furnace	Boron	[33]
3	Thermal	Furnace	Carbon dioxide	[34]
4	Thermal	Furnace	Carbon dioxide	[40]
5	Thermal	Furnace	Cyanide	[23]
6	Thermal	Furnace	Lead	[50]
7	Thermal	Furnace	Carbon dioxide	[58]
8	Thermal	Furnace	Phosphorus	[43]
9	Thermal	Furnace	Phenol	[44]
10	Thermal	Furnace	Phosphorous	[45]
11	Thermal	Furnace	Basic blue 9 Acid orange 51	[46]
12	Thermal	Furnace	Phosphate	[47]
13	Thermal	Furnace	Chromium	[48]
14	Thermal	Furnace	Cadmium Chromium Lead	[49]

15	Chemical	H_3PO_4 Calcium hydroxide	Cadmium Copper	[38]
16	Chemical	Iron oxide	Copper	[50]
17	Chemical	Methanol Hydrochloric acid	Boron	[33]
18	Chemical	$FeCl_3 \cdot 6H2O$ Potassium hydroxide sodium alginate calcium chloride	Congo red	[51]
19	Chemical	Hydrochloric acid (0.5%)	Cyanide	[52]
20	Chemical	Iron chloride	Phosphate	[25, 53]
21	Chemical	Magnesium nitrate Aluminium nitrate Urea Hydrochloric acid	Chromium	[53]
22	Chemical	Titanium tetraisopropoxide Hydrochloric acid Cadmium acetate dehydrate ammonium sulphide	Methylene blue	[24]
23	Chemical	5% polyvinyl alcohol 0.5% sodium alginate Calcium chloride Boric acid	Reactive red dye	[102]
24	Chemical	40 % acetic acid Ammonia nickel chloride	Copper	[55]
25	Chemical	Nitric acid Phosphoric acid Ammonium hydroxide	Nickel	[56]
26	Chemical	Phosphoric acid Calcium hydroxide	Lead	[26]
27	Thermal, chemical	goethite, a-MnO_2 and goethite/a-MnO_2	Arsenate	[57]

28	Thermal, chemical	Calcium chloride Sodium alginate Ammonia	Carbon dioxide	[58]
29	Thermal, chemical	Nitric acid (NH4)$_3$PO$_4$	Copper Zinc Lead	[59]
30	Thermal, chemical	Furnace, dimethylformamide (DMF)	Lead	[60]
31	Thermal, chemical	Furnace ferric sulphate	pathogenic bacteria and antibiotic resistance genes	[61]
32	Thermal, chemical	Furnace Sulphuric acid Sodium hydroxide Aluminium nitrate Aluminium sulphate.	Fluoride	[62]
33	Thermal, chemical	Furnace Nitric acid Sulphuric acid	Congo red	[63]
34	Thermal, chemical	Furnace Sodium hydroxide Hydrochloric acid Sodium citrate monohydrate, Sodium dihydrogen phosphate Sodium tetraborate Glycine Nitric acid	Lead	[64]
35	Physical	Ball mill	Cadmium	[100]

8. Eggshell waste for dye removal

Apart from heavy metals, dyes and pigments also pose an environmental threat as they are emitted into wastewaters from various industries such as dye manufacturing, textile finishing, food, cosmetics, paper and carpet industries [35]. The adsorption characteristics of basic yellow 28 dye onto calcined eggshells was investigated in batch studies which revealed a maximum biosorption capacity of 28.87 mg/g [35]. Results obtained inferred multilayer adsorption from the Freundlich model which was in agreement with work conducted by other researchers in the removal of brilliant green dye, methylene blue and congo red dye and reactive red 123 dye [29, 31, 96, 103]. Contrasting to these views, studies by Zulfikar et al. [30] involving the adsorption of congo red (CR) using untreated powdered eggshell, revealed the Langmuir model as the best fit model, as did Elkady et al. [54] who used immobilized eggshell with a polymer mixture of alginate and polyvinyl alcohol applied as a biocomposite adsorbent in the removal of C.I. Remazol Reactive Red 198 dye.

In investigating optimal conditions for dye removal, many studies have been conducted. An adsorption capacity of 1.26 mg/g was achieved in the removal of reactive red 123 dye by Ehrampoush et al. [103] with Kobiraj et al. [29] being able to achieve capacities of 44.7 mg/g, 34.23 mg/g and 30.23 mg/g at 303 K, 313 K and 323 K respectively in the removal of brilliant green dye [29]. Zulfikar et al.[30] studied the adsorption of congo red (CR) achieving a maximum capacity of 95.25 mg/g at a contact time of 20 minutes, adsorbent dosage of 20 g, initial concentration of 20 mg/L and pH 2. In elaborating on this, the maximum capacity of methylene blue and congo red was 94.9 mg/g and 49.5 mg/g for a concentration of 1000 mg/L at room temperature in a study by Abdel-Khalek et al. [96] who investigated the feasibility of the use of whole eggshell matrix (eggshell and membrane) to remove methylene blue and congo red. Mohamad et al. [104] on the other hand, focused on malachite green removal, achieving a 92.39% removal efficiency with optimal conditions of 70 mg/L, dosage of 1.99 g and contact time of 16.25 minutes using eggshell biochar.

Belay [97] used thermally treated eggshell to remove methyl orange and was able to achieve a 98.8% removal efficiency with conditions of 12.5mg/L, 2g adsorbent with a contact time of 20 minutes. Batch studies conducted by Elkady et al. [54] using immobilized eggshell with a polymer mixture of alginate and polyvinyl alcohol applied as a biocomposite adsorbent to remove C.I. Remazol Reactive Red 198 dye revealed a maximum dye removal capacity of 46.9 mg/g at the optimum pH 1.0 and 22 °C. An adsorption capacity of 1.26 mg/g was achieved when chicken eggshells were used to remove of reactive red 123 dye [31].

Podstawczyk et al.[105]focused on the removal mechanism of malachite green dye where it was found that physical adsorption, alkaline fading phenomenon and microprecipitation were the main mechanisms. Elaborating on this, Kobiraj et al.[29] found that adsorption followed both surface and intra-particle diffusion mechanisms in a study using hen eggshell powder to remove brilliant green dye [29].

189

Table 3. Biosorptive capacity of dyes using eggshell waste

No	Adsorbate	Adsorption capacity, $q_m \left(\frac{mg}{g}\right)$	Characteriza-tion	Isotherms	Kinetic models	Ref.
1	Basic yellow 28 dye	28.87	FTIR SEM BET XRD	Freundlich Tempkin Toth Dubinin Radushkevich Sips Koble-Corrigan	Pseudo 1st order pseudo 2nd order	[35]
2	Methylene blue	1	FTIR BET	Langmuir Freundlich		[21]
3	Reactive red dye	46.93	XRD TGA FTIR SEM	Langmuir Freundlich Tempkin	Pseudo 2nd order Intraparticle diffusion Boyd	[54]
4	Basic blue 9 Acid orange	113.6	FTIR BET	Langmuir Freundlich	Pseudo 2nd order	[46]
5	Brilliant green	44.7		Langmuir Freundlich Intraparticle diffusion	Pseudo 2nd order	[29]
6	Congo red	95.25		Langmuir Freundlich Sips	-	[30]
7	Methylene blue Congo red	94.9 49.5	FTIR FESEM	Langmuir Freundlich	Pseudo 1st order Pseudo 2nd order	[96]
8	Reactive Red 123	1.26	SEM EDX BET	Langmuir Freundlich Temkin	Pseudo 1st order Pseudo 2nd order	[31]

9	Congo red	136.99	SEM XRD Zeta potential	Langmuir Freundlich Tempkin Dubinin Raduskovich	Pseudo 1st order pseudo 2nd order Elovich Intraparticle diffusion	[63]
10	Rhodamine B Eriochrome black T Murexide	1.20 1.03 1.57	FTIR TGA SEM XRD	Langmuir Freundlich Tempkin Dubinin Raduskovich	Pseudo 1st order pseudo 2nd order	[106]
11	Styrylpyridin ium dye	166.67	FTIR SEM	Langmuir Freundlich	Pseudo 1st order pseudo 2nd order	[107]
12	Reactive red 35	41.85	BET FTIR	Langmuir Freundlich	Pseudo 1st order pseudo 2nd order Intraparticle diffusion	[108]

Concluding remarks

Conventional technologies used to remove pollutants from waste effluent heavily rely on expensive chemicals for the treatment of industrial waste effluent. However, due to costs, technical applicability and stringent environmental regulations, the use of locally available biomass such as eggshell waste is an attractive alternative. From the literature cited in this review, eggshell waste is an excellent source of inorganic and bio-organic materials. Suitable processing may be developed for use as a biosorbent in the water treatment industry. Due to their micro-porous structure they behave as a good biosorbent for removing various contaminants such as textile dyes and heavy metals amongst other pollutants. Since this material is abundant, low-cost and biodegradable, it can be employed in the process of biosorption.

It was found that the efficiency of a biosorption process is governed by the form of biomass present, which can be used in its natural form, modified or its derivatives. The mechanism of biosorption is often related to its process chemistry which determines the efficacy of the biomass used. In most cases, the Langmuir isotherm model can be used to adequately describe the interaction between the adsorbate and adsorbent which is indicative that biosorption occurs as a monolayer on a homogenous surface. In addition, the pseudo 2nd order kinetic modelling equation works well in the description and prediction of metal uptake. In terms of surface functional groups, carbonate groups mainly consisting of carbon and oxygen play a key role in the uptake of metals and byes by the biomass.

Research has provided a better understanding of biosorption to a certain extent, however, there are shortcomings which cannot be overlooked such as competitive adsorption which has been limited in its study. Additionally, real wastewater interactions between the adsorbate-adsorbent systems have not been adequately investigated and further research into this is needed. On the contrary, eggshell waste can be considered a viable potential low-cost biomass however more information is required to determine the best combination of pollutants that can be removed which is highly dependent on environmental conditions.

Abbreviations

FTIR   Fourier Transform Infrared Spectroscopy

SEM   Scanning Electron Microscopy

XRD   X-Ray Diffraction

EDX   Energy Dispersive Spectroscopy

TEM   Transmission Electron Microscopy

BET   Brunauer Emmett Teller

FESEM   Field Emission Scanning Electron Microscopy

TGA   Thermal Gravimetric Analysis

References

[1] A. Abdolali, W. S. Guo, H. H. Ngo, S. S. Chen, N. C. Nguyen, and K. L. Tung, Typical lignocellulosic wastes and by-products for biosorption process in water and wastewater treatment: a critical review, Bioresource technology, vol. 160, pp. 57-66, 2014. https://doi.org/10.1016/j.biortech.2013.12.037

[2] S. Senthikumaar, Bharathi, S., Nithyanandhi, D., and Subburaam, C.V. , Biosorption of toxic heavy metals from aqueous solutions Bioresource Technology, vol. 75, pp. 163-165, 2000. https://doi.org/10.1016/S0960-8524(00)00021-3

Advances in Wastewater Treatment I   Materials Research Forum LLC
Materials Research Foundations 91 (2021) 171-   https://doi.org/10.21741/9781644901144-5

[3] D. Lakherwal, Adsorption of Heavy Metals: A Review International Journal of Environmental Research and Development, vol. 4, no. 41-48, 2014.

[4] A. Ronda, M. A. Martín-Lara, E. Dionisio, G. Blázquez, and M. Calero, Effect of lead in biosorption of copper by almond shell, Journal of the Taiwan Institute of Chemical Engineers, vol. 44, no. 3, pp. 466-473, 2013. https://doi.org/10.1016/j.jtice.2012.12.019

[5] A. Luptakova, S. Ubaldini, P. Fornari, and E. Macingova, Physical, chemical l chemical methods for treatment of Acid mine drainage, Journal of Chemical Engineering Transactions, 2012.

[6] R. W. Gaikwad, V. S. S. Sapkal, and R. S. S. Sapkal, Ion exchange system design for removal of heavy metals from acid mine drainage wastewater, Acta Montanistica Slovaca, pp. 298-304, 2010.

[7] D. C. Buzzi, L. S. Viegas, F. P. C. Silvas, and D. C. R. Espinosa, The use of Microfiltration and electrodialysis for Treatment of Acid Mine drainage, IMWA, pp. 287-292, 2011.

[8] G. S. Simate and S. Ndlovu, The removal of heavy metals in a packed bed column using immobilized cassava peel waste biomass, Journal of Industrial and Engineering Chemistry, vol. 21, pp. 635-643, 2015. https://doi.org/10.1016/j.jiec.2014.03.031

[9] Z. Aksu and F. Gönen, Biosorption of phenol by immobilized activated sludge in a continuous packed bed: prediction of breakthrough curves, Process Biochem., vol. 5, no. 39, pp. 599-613, 2004.

[10] E. Malkoc, Y. Nuhoglu, and Y. Abali, Cr(VI) adsorption by waste acorn of Quercus ithaburensis in fixed beds: Prediction of breakthrough curves, Chemical Engineering Journal, vol. 119, no. 1, pp. 61-68, 2006. https://doi.org/10.1016/j.cej.2006.01.019

[11] South African Poultry Association

[12] SAPA. (2015). South African Poultry Association. Available: www.sapoultry.co.za

[13] D. Oliveira, P. Benelli, and E. Amante, A literature review on adding value to solid residues: Egg shells. 2013, pp. 42-47. https://doi.org/10.1016/j.jclepro.2012.09.045

[14] H. J. a. L. Choi , S.M., Heavy metal removal from acid mine drainage by calcined eggshell and microalgae hybrid system, Environmental Science and Pollution Research, vol. 22, pp. 13404-13411, 2015. https://doi.org/10.1007/s11356-015-4623-3

[15] J. Carvalho, A. Ribeiro, J. Graça, J. Araújo, C. Vilarinho, and F. Castro, Adsorption Process Onto An Innovative Eggshell-Derived Low-Cost Adsorbent In Simulated Effluent And Real Industrial Effluents. 2011.

[16] A. King ori, A Review of the uses of poultry eggshells and shell membranes. 2011, pp. 908-912. https://doi.org/10.3923/ijps.2011.908.912

[17] Z. Zhang, A. M Gonzalez, E. Davies, and Y. Liu, Agricultural Wastes. 2012, pp. 1386-1406. https://doi.org/10.2175/106143012X13407275695193

[18] J. V. Flores-Cano, R. Leyva-Ramos, J. Mendoza-Barron, R. M. Guerrero-Coronado, A. Aragón-Piña, and G. J. Labrada-Delgado, Sorption mechanism of Cd(II) from water solution onto chicken eggshell, Applied Surface Science, vol. 276, pp. 682-690, 2013/07/01/ 2013. https://doi.org/10.1016/j.apsusc.2013.03.153

[19] M. T. Hincke, Y. Nys, J. Gautron, K. Mann, and A. B. Rodriguez-Navarro, The eggshell: Structure, composition and mineralization., Front. Biosci. Special Edition on Biomineralization, vol. 17, pp. 1266-1280, 2012. https://doi.org/10.2741/3985

[20] T. Nakano, N. I Ikawa, and L. Ozimek, Chemical composition of chicken eggshell and shell membranes. 2003, pp. 510-4. https://doi.org/10.1093/ps/82.3.510

[21] W. T. Tsai, J. M. Yang, C. W. Lai, Y. H. Cheng, C. C. Lin, and C. W. Yeh, Characterization and adsorption properties of eggshells and eggshell membrane, Bioresource Technology, vol. 97, no. 3, pp. 488-493, 2006/02/01/ 2006. https://doi.org/10.1016/j.biortech.2005.02.050

[22] A. Mittal, M. Teotia, R. K. Soni, and J. Mittal, Applications of egg shell and egg shell membrane as adsorbents: A review, Journal of Molecular Liquids, vol. 223, pp. 376-387, 2016/11/01/ 2016. https://doi.org/10.1016/j.molliq.2016.08.065

[23] O. A. A. Eletta, O. A. Ajayi, O. O. Ogunleye, and I. C. Akpan, Adsorption of cyanide from aqueous solution using calcinated eggshells: Equilibrium and optimisation studies, Journal of Environmental Chemical Engineering, vol. 4, no. 1, pp. 1367-1375, 2016/03/01/ 2016. https://doi.org/10.1016/j.jece.2016.01.020

[24] B. Pant, M. Park, H.-Y. Kim, and S.-J. Park, CdS-TiO2 NPs decorated carbonized eggshell membrane for effective removal of organic pollutants: A novel strategy to use a waste material for environmental remediation, Journal of Alloys and Compounds, vol. 699, pp. 73-78, 2017/03/30/ 2017. https://doi.org/10.1016/j.jallcom.2016.12.360

[25] N. Y. Mezenner and A. Bensmaili, Kinetics and thermodynamic study of phosphate adsorption on iron hydroxide-eggshell waste, Chemical Engineering Journal, vol. 147, no. 2, pp. 87-96, 2009/04/15/ 2009. https://doi.org/10.1016/j.cej.2008.06.024

[26] D. Liao et al., Removal of lead(II) from aqueous solutions using carbonate hydroxyapatite extracted from eggshell waste, Journal of Hazardous Materials, vol. 126-130, 2010/05/15/ 2010. https://doi.org/10.1016/j.jhazmat.2009.12.005

[27] O. Habeeb, R. Kanthasamy, G. Ali, R. Yunus, and O. Olalere, Kinetic, Isotherm and Equilibrium Study of Adsorption of Hydrogen Sulfide From Wastewater Using Modified Eggshells. 2017, pp. 13-25. https://doi.org/10.31436/iiumej.v18i1.689

[28] K. Chojnacka, Biosorption of Cr(III) ions by eggshells, Journal of Hazardous Materials, vol. 121, no. 1, pp. 167-173, 2005/05/20/ 2005. https://doi.org/10.1016/j.jhazmat.2005.02.004

[29] R. Kobiraj, N. Gupta, A. K. Kushwaha, and M. C. Chattopadhyaya, Determination of equilibrium, kinetic and thermodynamic parameters for the adsorption of Brilliant Green dye from aqueous solutions onto eggshell powder, Indian Journal of Chemical Technology, vol. 19, pp. 26-31, 2012.

[30] M. A. Zulfikar and H. Setiyanto, Study of the adsorption kinetics and thermodynamic for the removal of Congo Red from aqueous solution using powdered eggshell. 2013, pp. 1671-1678.

[31] M. H. Ehrampoush, G. Ghanizadeh, and M. H. Ghaneian, Equilibrium And Kinetics Study Of Reactive Red 123 Dye Removal From Aqueous Solution By Adsorption On Eggshell Iran. J. Environ. Health. Sci. Eng, vol. 8, no. 2, pp. 101-108, 2011.

[32] R. Slimani et al., Calcined eggshells as a new biosorbent to remove basic dye from aqueous solutions: Thermodynamics, kinetics, Isotherms and error analysis, Journal of the Taiwan Institute of Chemical Engineers, vol. 45, no. 4, pp. 1578-1587, 2014/07/01/ 2014. https://doi.org/10.1016/j.jtice.2013.10.009

[33] M. A. Al-Ghouti and N. R. Salih, Application of eggshell wastes for boron remediation from water, Journal of Molecular Liquids, vol. 256, pp. 599-610, 2018/04/15/ 2018. https://doi.org/10.1016/j.molliq.2018.02.074

[34] T. Witoon, Characterization of calcium oxide derived from waste eggshell and its application as CO2 sorbent, Ceramics International, vol. 37, no. 8, pp. 3291-3298, 2011/12/01/ 2011. https://doi.org/10.1016/j.ceramint.2011.05.125

[35] R. Slimani et al., Calcined eggshells as a new biosorbent to remove basic dye from aqueous solutions: Thermodynamics, kinetics, isotherms and error analysis, Journal Of The Taiwan Institute Of Chemical Engineers, vol. 45, no. 4, pp. 1578-1587, 2014. https://doi.org/10.1016/j.jtice.2013.10.009

[36] M. Baláž, A. Zorkovská, M. Fabián, V. Girman, and J. Briančin, Eggshell biomaterial: Characterization of nanophase and polymorphs after mechanical activation, Advanced Powder Technology, vol. 26, no. 6, pp. 1597-1608, 2015/11/01/ 2015. https://doi.org/10.1016/j.apt.2015.09.003

[37] M. Baláž, J. Ficeriová, and J. Briančin, Influence of milling on the adsorption ability of eggshell waste, Chemosphere, vol. 146, pp. 458-471, 2016/03/01/ 2016. https://doi.org/10.1016/j.chemosphere.2015.12.002

[38] W. Zheng et al., Adsorption of Cd(II) and Cu(II) from aqueous solution by carbonate hydroxylapatite derived from eggshell waste, Journal of Hazardous Materials, vol. 147, no. 1, pp. 534-539, 2007/08/17/ 2007. https://doi.org/10.1016/j.jhazmat.2007.01.048

[39] S. Ramesh et al., Direct conversion of eggshell to hydroxyapatite ceramic by a sintering method, Ceramics International, vol. 42, no. 6, pp. 7824-7829, 2016/05/01/ 2016. https://doi.org/10.1016/j.ceramint.2016.02.015

[40] S. Shan, Ma, A., Hu ,Y., Jia ,Q., Wang, Y. and Peng, J., Development of sintering-resistant CaO-based sorbent derived from eggshells and bauxite tailings for cyclic CO2 capture, Environmental Pollution, no. 208, pp. 546-552, 2016. https://doi.org/10.1016/j.envpol.2015.10.028

[41] M. Ahmad, Y. Hashimoto, D. H. Moon, S. S. Lee, and Y. S. Ok, Immobilization of lead in a Korean military shooting range soil using eggshell waste: An integrated mechanistic approach, Journal of Hazardous Materials, vol. 209-210, pp. 392-401, 2012/03/30/ 2012. https://doi.org/10.1016/j.jhazmat.2012.01.047

[42] S. He et al., Investigation of CaO-based sorbents derived from eggshells and red mud for CO2 capture, Journal of Alloys and Compounds, vol. 701, pp. 828-833, 2017/04/15/ 2017. https://doi.org/10.1016/j.jallcom.2016.12.194

[43] M. Oliveira, A. Araújo, G. Azevedo, M. F. R. Pereira, I. C. Neves, and A. V. Machado, Kinetic and equilibrium studies of phosphorous adsorption: Effect of physical and chemical properties of adsorption agent, Ecological Engineering, vol. 82, pp. 527-530, 2015/09/01/ 2015. https://doi.org/10.1016/j.ecoleng.2015.05.020

[44] L. Giraldo and J. C. Moreno-Piraján, Study of adsorption of phenol on activated carbons obtained from eggshells, Journal of Analytical and Applied Pyrolysis, vol. 106, pp. 41-47, 2014/03/01/ 2014. https://doi.org/10.1016/j.jaap.2013.12.007

[45] E. Panagiotou et al., Turning calcined waste egg shells and wastewater to Brushite: Phosphorus adsorption from aqua media and anaerobic sludge leach water, Journal of

Cleaner Production, vol. 178, pp. 419-428, 2018/03/20/ 2018.
https://doi.org/10.1016/j.jclepro.2018.01.014

[46] W.-T. Tsai, K.-J. Hsien, H.-C. Hsu, C.-M. Lin, K.-Y. Lin, and C.-H. Chiu,
Utilization of ground eggshell waste as an adsorbent for the removal of dyes from
aqueous solution, Bioresource Technology, vol. 99, no. 6, pp. 1623-1629, 2008/04/01/
2008. https://doi.org/10.1016/j.biortech.2007.04.010

[47] T. E. Köse and B. Kıvanç, Adsorption of phosphate from aqueous solutions using
calcined waste eggshell, Chemical Engineering Journal, vol. 178, pp. 34-39,
2011/12/15/ 2011. https://doi.org/10.1016/j.cej.2011.09.129

[48] M. A. Renu, K. Singh, S. Upadhyaya, and R. K. Dohare, Removal of heavy metals
from wastewater using modified agricultural adsorbents, Materials Today:
Proceedings, vol. 4, no. 9, pp. 10534-10538, 2017/01/01/ 2017.
https://doi.org/10.1016/j.matpr.2017.06.415

[49] H. J. Park, S. W. Jeong, J. K. Yang, B. G. Kim, and S. M. Lee, Removal of heavy
metals using waste eggshell, Journal of Environmental Sciences, vol. 19, no. 12, pp.
1436-1441, 2007/01/01/ 2007. https://doi.org/10.1016/S1001-0742(07)60234-4

[50] M. Ahmad, Usman, A.R.A., Lee, S.S., Kim, S.C., Joo, J.H., Yang, J.E. and Ok, Y.S.,
Eggshell and coral wastes as low cost sorbents for the removal of Pb2+, Cd2+ and
Cu2+ from aqueous solutions, Journal Of Industrial And Engineering Chemistry, vol.
18, no. 1, pp. 198-204, 2012. https://doi.org/10.1016/j.jiec.2011.11.013

[51] V. S. Munagapati and D.-S. Kim, Equilibrium isotherms, kinetics, and
thermodynamics studies for congo red adsorption using calcium alginate beads
impregnated with nano-goethite, Ecotoxicology and Environmental Safety, vol. 141,
pp. 226-234, 2017/07/01/ 2017. https://doi.org/10.1016/j.ecoenv.2017.03.036

[52] G. Asgari and A. Dayari, Experimental dataset on acid treated eggshell for removing
cyanide ions from synthetic and industrial wastewaters, Data in Brief, vol. 16, pp. 442-
452, 2018/02/01/ 2018. https://doi.org/10.1016/j.dib.2017.11.048

[53] X. Guo et al., Layered double hydroxide/eggshell membrane: An inorganic
biocomposite membrane as an efficient adsorbent for Cr(VI) removal, Chemical
Engineering Journal, vol. 166, no. 1, pp. 81-87, 2011/01/01/ 2011.
https://doi.org/10.1016/j.cej.2010.10.010

[54] M. Elkady, A. Ibrahim, and M. Abd El-Latif, Assessment of the adsorption kinetics,
equilibrium and thermodynamic for the potential removal of reactive red dye using

eggshell biocomposite beads. 2011, pp. 412-423.
https://doi.org/10.1016/j.desal.2011.05.063

[55] J. H. Chen et al., Recovery and investigation of Cu(II) ions by tannin immobilized porous membrane adsorbent from aqueous solution, Chemical Engineering Journal, vol. 273, pp. 19-27, 2015/08/01/ 2015. https://doi.org/10.1016/j.cej.2015.03.031

[56] G. De Angelis, L. Medeghini, A. M. Conte, and S. Mignardi, Recycling of eggshell waste into low-cost adsorbent for Ni removal from wastewater, Journal of Cleaner Production, vol. 164, pp. 1497-1506, 2017/10/15/ 2017. https://doi.org/10.1016/j.jclepro.2017.07.085

[57] J. S. Markovski et al., Arsenate adsorption on waste eggshell modified by goethite, α-MnO2 and goethite/α-MnO2, Chemical Engineering Journal, vol. 237, pp. 430-442, 2014/02/01/ 2014. https://doi.org/10.1016/j.cej.2013.10.031

[58] S. Hosseini, F. Eghbali Babadi, S. Masoudi Soltani, M. K. Aroua, S. Babamohammadi, and A. Mousavi Moghadam, Carbon dioxide adsorption on nitrogen-enriched gel beads from calcined eggshell/sodium alginate natural composite, Process Safety and Environmental Protection, vol. 109, pp. 387-399, 2017/07/01/ 2017. https://doi.org/10.1016/j.psep.2017.03.021

[59] W. I. Mortada, I. M. M. Kenawy, A. M. Abdelghany, A. M. Ismail, A. F. Donia, and K. A. Nabieh, Determination of Cu2+, Zn2+ and Pb2+ in biological and food samples by FAAS after preconcentration with hydroxyapatite nanorods originated from eggshell, Materials Science and Engineering: C, vol. 52, pp. 288-296, 2015/07/01/ 2015. https://doi.org/10.1016/j.msec.2015.03.061

[60] H. Wang et al., Engineered biochar derived from eggshell-treated biomass for removal of aqueous lead, Ecological Engineering, 2017/07/08/ 2017.

[61] Y. Mao, Mingming, S., Xu, C., Yanfang, F., Jinzhong, W., Kuan, L., Da, T., Manqiang, L., Jun, W., Schwab, A.P., and Xin, J., Feasibility of sulfate-calcined eggshells for removing pathogenic bacteria and antibiotic resistance genes from landfill leachates, Waste Management, vol. 63, pp. 275-283, 2017/05/01/ 2017. https://doi.org/10.1016/j.wasman.2017.03.005

[62] S. Lunge, D. Thakre, S. Kamble, N. Labhsetwar, and S. Rayalu, Alumina supported carbon composite material with exceptionally high defluoridation property from eggshell waste, Journal of Hazardous Materials, vol. 237-238, pp. 161-169, 2012/10/30/ 2012. https://doi.org/10.1016/j.jhazmat.2012.08.023

[63] E. N. Seyahmazegi, R. Mohammad-Rezaei, and H. Razmi, Multiwall carbon nanotubes decorated on calcined eggshell waste as a novel nano-sorbent: Application f or anionic dye Congo red removal, Chemical Engineering Research and Design, vol. 109, pp. 824-834, 2016/05/01/ 2016. https://doi.org/10.1016/j.cherd.2016.04.001

[64] O. Kaplan Ince, M. Ince, V. Yonten, and A. Goksu, A food waste utilization study for removing lead(II) from drinks, Food Chemistry, vol. 214, pp. 637-643, 2017/01/01/ 2017. https://doi.org/10.1016/j.foodchem.2016.07.117

[65] A. E. Burakov et al., Adsorption of heavy metals on conventional and nanostructured materials for wastewater treatment purposes: A review, Ecotoxicology and Environmental Safety, vol. 148, pp. 702-712, 2018/02/01/ 2018. https://doi.org/10.1016/j.ecoenv.2017.11.034

[66] I. Anastopoulos and G. Z. Kyzas, Agricultural peels for dye adsorption: A review of recent literature, Journal of Molecular Liquids, vol. 200, pp. 381-389, 2014/12/01/ 2014. https://doi.org/10.1016/j.molliq.2014.11.006

[67] Y. Jin et al., Batch and fixed-bed biosorption of Cd(II) from aqueous solution using immobilized Pleurotus ostreatus spent substrate, Chemosphere, vol. 191, pp. 799-808, 2018/01/01/ 2018. https://doi.org/10.1016/j.chemosphere.2017.08.154

[68] I. Morosanu, C. Teodosiu, C. Paduraru, D. Ibanescu, and L. Tofan, Biosorption of lead ions from aqueous effluents by rapeseed biomass, NEW BIOTECHNOLOGY, vol. 39, pp. 110-124, 2017. https://doi.org/10.1016/j.nbt.2016.08.002

[69] C. E. R. Barquilha, E. S. Cossich, C. R. G. Tavares, and E. A. Silva, Biosorption of nickel(II) and copper(II) ions in batch and fixed-bed columns by free and immobilized marine algae Sargassum sp, Journal Of Cleaner Production, vol. 150, pp. 58-64, 2017. https://doi.org/10.1016/j.jclepro.2017.02.199

[70] S. Guiza, Biosorption of heavy metal from aqueous solution using cellulosic waste orange peel, Ecological Engineering, vol. 99, pp. 134-140, 2017/02/01/ 2017. https://doi.org/10.1016/j.ecoleng.2016.11.043

[71] K. Y. Foo and B. H. Hameed, An overview of dye removal via activated carbon adsorption process, Desalination and Water Treatment, vol. 19, no. 1-3, pp. 255-274, 2010/07/01 2010. https://doi.org/10.5004/dwt.2010.1214

[72] S. Gautam, P. Kumar, and A. Patra, Gautam et al 2014. 2015.

[73] S. Rangabhashiyam, N. Anu, M. S. Giri Nandagopal, and N. Selvaraju, Relevance of isotherm models in biosorption of pollutants by agricultural byproducts, Journal of

Materials Research Forum LLC
https://doi.org/10.21741/9781644901144-5

Environmental Chemical Engineering, vol. 2, no. 1, pp. 398-414, 2014/03/01/ 2014. https://doi.org/10.1016/j.jece.2014.01.014

[74] G. Blázquez, M. Calero, A. Ronda, G. Tenorio, and M. A. Martín-Lara, Study of kinetics in the biosorption of lead onto native and chemically treated olive stone, Journal of Industrial and Engineering Chemistry, vol. 20, no. 5, pp. 2754-2760, 2014/09/25/ 2014. https://doi.org/10.1016/j.jiec.2013.11.003

[75] S. Moussous, A. Selatnia, A. Merati, and G. A. junter, Batch cadmium(II) biosorption by an industrial residue of macrofungal biomass (Clitopilus scyphoides), Chemical Engineering Journal, vol. 197, pp. 261-271, 2012/07/15/ 2012. https://doi.org/10.1016/j.cej.2012.04.106

[76] N. Barka, M. Abdennouri, M. El Makhfouk, and S. Qourzal, Biosorption characteristics of cadmium and lead onto eco-friendly dried cactus (Opuntia ficus indica) cladodes, Journal of Environmental Chemical Engineering, vol. 1, no. 3, pp. 144-149, 2013/09/01/ 2013. https://doi.org/10.1016/j.jece.2013.04.008

[77] L. Pelit, F. N. Ertaş, A. E. Eroğlu, T. Shahwan, and H. Tural, Biosorption of Cu(II) and Pb(II) ions from aqueous solution by natural spider silk, Bioresource Technology, vol. 102, no. 19, pp. 8807-8813, 2011/10/01/ 2011. https://doi.org/10.1016/j.biortech.2011.07.013

[78] A. Witek-Krowiak, Analysis of influence of process conditions on kinetics of malachite green biosorption onto beech sawdust, Chemical Engineering Journal, vol. 171, no. 3, pp. 976-985, 2011/07/15/ 2011. https://doi.org/10.1016/j.cej.2011.04.048

[79] S. Qaiser, A. R. Saleemi, and M. Umar, Biosorption of lead from aqueous solution by Ficus religiosa leaves: Batch and column study, Journal of Hazardous Materials, vol. 166, no. 2, pp. 998-1005, 2009. https://doi.org/10.1016/j.jhazmat.2008.12.003

[80] A. Verma, S. Kumar, and S. Kumar, Biosorption of lead ions from the aqueous solution by Sargassum filipendula: Equilibrium and kinetic studies, Journal of Environmental Chemical Engineering, vol. 4, no. 4, pp. 4587-4599, 2016. https://doi.org/10.1016/j.jece.2016.10.026

[81] J. Park, S. W. Won, J. Mao, I. S. Kwak, and Y.-S. Yun, Recovery of Pd(II) from hydrochloric solution using polyallylamine hydrochloride-modified Escherichia coli biomass, Journal of Hazardous Materials, vol. 181, no. 1, pp. 794-800, 2010/09/15/ 2010. https://doi.org/10.1016/j.jhazmat.2010.05.083

[82] F. Veglio and F. Beolchini, Removal of metals by biosorption: a review, Hydrometallurgy, vol. 44, no. 3, pp. 301-316, 1997/03/01/ 1997. https://doi.org/10.1016/S0304-386X(96)00059-X

[83] W. P. Putra et al., Biosorption of Cu(II), Pb(II) and Zn(II) Ions from Aqueous Solutions Using Selected Waste Materials: Adsorption and Characterisation Studies, Journal of Encapsulation and Adsorption Sciences, vol. Vol.04No.01, p. 11, 2014, Art. no. 43532. https://doi.org/10.4236/jeas.2014.41004

[84] B. M. W. P. K. Amarasinghe and R. A. Williams, Tea Waste as a Low Cost Adsorbent for The Removal of Cu and Pb from Wastewater. 2007, pp. 299-309. https://doi.org/10.1016/j.cej.2007.01.016

[85] A. Schaafsma, I. Pakan, G. J. H. Hofstede, F. A. Muskiet, E. Van Der Veer, and P. J. F. De Vries, Mineral, amino acid, and hormonal composition of chicken eggshell powder and the evaluation of its use in human nutrition, Poultry Science, vol. 79, no. 12, pp. 1833-1838, 2000. https://doi.org/10.1093/ps/79.12.1833

[86] K. Vijayaraghavan, J, and M. Umid Chicken Eggshells Remove Pb(II) Ions from Synthetic Wastewater, Environmental Engineering Science, vol. 30, no. 2, pp. 67-73, 2013. https://doi.org/10.1089/ees.2012.0038

[87] N. Das, Recovery of precious metals through biosorption - A review, Hydrometallurgy, vol. 103, no. 1, pp. 180-189, 2010/06/01/ 2010. https://doi.org/10.1016/j.hydromet.2010.03.016

[88] R. Gao and J. Wang, Effects of pH and temperature on isotherm parameters of chlorophenols biosorption to anaerobic granular sludge, Journal of Hazardous Materials, vol. 145, no. 3, pp. 398-403, 2007/07/16/ 2007. https://doi.org/10.1016/j.jhazmat.2006.11.036

[89] U. Farooq, J. A. Kozinski, M. A. Khan, and M. Athar, Biosorption of heavy metal ions using wheat based biosorbents - A review of the recent literature, Bioresource Technology, vol. 101, no. 14, pp. 5043-5053, 2010. https://doi.org/10.1016/j.biortech.2010.02.030

[90] K. C. Malakondaiah, D. A. Kalpana, D. A. Naidu, P. King, and V. S. R. K. Prasad, Low cost biosorbent for the removal of Cu (II) from aqueous solution, Journal of Environmental Science, vol. 5, pp. 363-368, 2010.

[91] J. He and J. P. Chen, A comprehensive review on biosorption of heavy metals by algal biomass: Materials, performances, chemistry, and modeling simulation tools,

Bioresource Technology, vol. 160, pp. 67-78, 2014/05/01/ 2014. https://doi.org/10.1016/j.biortech.2014.01.068

[92] D. Senthilkumar, Ethiraj, A.S., Vimala, R., Ramalingam, C. and Jayanthi, S., Biosorption of Cu (II) from aqueous solutions: Kinetics and characterization studies, Scholars Research Library, vol. 7, no. 3, pp. 205-213, 2015.

[93] Z. Tark, M. Ibrahim, and H. Madhloom, Eggshell Powder As An Adsorbent for Removal of Cu (II) and Cd (II) from Aqueous Solution: Equilibrium, Kinetic and Thermodynamic Studies. 2016, pp. 186-193.

[94] R. Bhaumik, Mondal, N.K., Das ,B., Roy, P., Pal, K.C., Das,C., Banerjee, A. And Datta< J.K., Eggshell Powder as an Adsorbent for Removal of Fluoride from Aqueous Solution: Equilibrium, Kinetic and Thermodynamic Studies, Journal of Chemistry, vol. 9, pp. 1457-1480, 2012. https://doi.org/10.1155/2012/790401

[95] N. Yeddou and A. Bensmaili, Equilibrium and kinetic modelling of iron adsorption by eggshells in a batch system: effect of temperature, Desalination, vol. 206, no. 1, pp. 127-134, 2007/02/05/ 2007. https://doi.org/10.1016/j.desal.2006.04.052

[96] M. A. Abdel-Khalek, M. K. Abdel Rahman, and A. A. Francis, Exploring the adsorption behavior of cationic and anionic dyes on industrial waste shells of egg, Journal of Environmental Chemical Engineering, vol. 5, no. 1, pp. 319-327, 2017/02/01/ 2017. https://doi.org/10.1016/j.jece.2016.11.043

[97] K. Belay, Removal of Methyl Orange from Aqueous Solutions Using Thermally Treated Egg Shell (Locally Available and Low Cost Biosorbent). 2015.

[98] M. a. B. Kanyal, A.A., Removal of Heavy Metals from Water (Cu and Pb) Using Household Waste as an Adsorbent. 2015.

[99] T. D. Mashangwa, Tekere, M. and Sibanda, T., Determination of the Efficacy of Eggshell as a Low-Cost Adsorbent for the Treatment of Metal Laden Effluents, Int Journal of Environ Res, vol. 11, pp. 175-188, 2017. https://doi.org/10.1007/s41742-017-0017-3

[100] M. Baláž, Z. Bujňáková, P. Baláž, A. Zorkovská, Z. Danková, and J. Briančin, Adsorption of cadmium(II) on waste biomaterial, Journal of Colloid and Interface Science, vol. 454, pp. 121-133, 2015/09/15/ 2015. https://doi.org/10.1016/j.jcis.2015.03.046

[101] Y. S. Ok, S. S. Lee, W.-T. Jeon, S.-E. Oh, A. Usman, and D. H. Moon, Application of Eggshell Waste for the Immobilization of Cadmium and Lead in a Contaminated

Soil, Environ Geochem Health, vol. 33 Suppl 1, pp. 31-9, 2011.
https://doi.org/10.1007/s10653-010-9362-2

[102] M. F. Elkady, A. M. Ibrahim, and M. M. A. El-Latif, Assessment of the adsorption
kinetics, equilibrium and thermodynamic for the potential removal of reactive red dye
using eggshell biocomposite beads, Desalination, vol. 278, no. 1, pp. 412-423,
2011/09/01/ 2011. https://doi.org/10.1016/j.desal.2011.05.063

[103] M. H. Ehrampoush, G. Ghanizadeh, and M. T. Ghaneian, Equilibrium And Kinetics
Study Of Reactive Red 123 Dye Removal From Aqueous Solution By Adsorption On
Eggshell. 2012, pp. 101-108.

[104] M. Mohamad, Wei, T.C., Mohammad, R. and Wei, L.J., Optimization Of Operating
Parameters By Responsesurface Methodology For Malachite Green Dye Removal
Using Biochar Prepared From Eggshell, Journal of Engineering and Applied Sciences
vol. 12, no. 11, pp. 3621-3633, 2017.

[105] D. Podstawczyk, A. Witek-Krowiak, K. Chojnacka, and Z. Sadowski, Biosorption
of malachite green by eggshells: Mechanism identification and process optimization,
Bioresource Technology, vol. 160, pp. 161-165, 2014/05/01/ 2014.
https://doi.org/10.1016/j.biortech.2014.01.015

[106] A. V. Borhade and A. S. Kale, Calcined eggshell as a cost effective material for
removal of dyes from aqueous solution, Applied Water Science, vol. 7, no. 8, pp.
4255-4268, 2017/12/01 2017. https://doi.org/10.1007/s13201-017-0558-9

[107] P. S. Guru and S. Dash, Amino Acid Modified Eggshell Powder (AA-ESP)-A
Novel Bio-Solid Scaffold for Adsorption of Some Styrylpyridinium Dyes, Journal of
Dispersion Science & Technology, Article vol. 34, no. 8, pp. 1099-1112, 2013.
https://doi.org/10.1080/01932691.2012.737752

[108] A. a. V. Babuponnusami, S., Investigation on adsorption of dye (Reactive Red 35)
on Egg shell powder, International Journal of ChemTech Research vol. 10, pp. 565-
572 2017.

Advances in Wastewater Treatment I

Materials Research Foundations **91** (2021) 204-218

Materials Research Forum LLC

https://doi.org/10.21741/9781644901144-6

Chapter 6

Removal of PAHs from Wastewater Using Powdered Activated Carbon: A Case Study

Zhaoyang You, Dongjian Cai, Jiaqing Tao, Kinjal J. Shah [*], Haiyang Xu

College of Urban Construction, Nanjing Tech University, Nanjing, 211800, China

kjshah@njtech.edu.cn

Abstract

The aim of this study was to develop immobilized microorganism carrier for effectively degradation of petroleum hydrocarbons (PAHs), especially pyrene. Powdered activated carbon (PAC) was used to immobilize the bacterial consortium (*Klebsiella pneumoniae* and *Pseudomonas aeruginosa*) with binder $CaCl_2$ and sodium alginate (SA) for improving mass transfer rate of the pyrene pollutants. Mass transfer properties, embedding ratio, and mechanical strength were inspected for the immobilization particles. Mechanical strength of SA beads was more influenced by proportion of SA and $CaCl_2$ than by proportion of PAC. The optimum proportion of SA, $CaCl_2$ and PAC were 2.5%, 2% and 0.5% for immobilization SA beads. The degradation of bacterial consortium (Pa+Kp) had the best degradation rates at 48.2% on 14 days. SA embedding immobilization by adding PAC can obviously enhanced effect of pyrene degradation because of bacterial absorption ability and nutrient permeability being improved.

Keywords

Petroleum Hydrocarbons, Powdered Activated Carbon, Mass Transfer Properties, Pyrene, Degradation

Contents

1.   Introduction

Water quality is the biggest challenges that people will face during the 21st century, threatening to human health and limiting food quality [1]. To assure the quality of water, Sustainable Development Goal has announced goal 6 as availability and sustainable management of water and sanitation for all [2]. In recent years, organic compounds such as polycyclic aromatic hydrocarbons (PAHs) have been commonly observed in water environment [3]. A recent study has stated the occurrence of PAHs in influent and effluent plant, groundwater, and surface and sea water [4]. The European Environment Agency and United State Environmental Protection Agency has reported the leakages of PAHs from petroleum industries leads to increase groundwater pollution [5]. Hence, it is essential to develop economic and viable way to limit emission or remove PAHs from water system to maintain health of ecosystem and human health.

Various physico-chemical and biological treatments have been used to limit the PAHs pollution [5]. However, physico-chemical processes are high in operation cost and generate secondary pollutants [6]. Thus, bioremediation treatments have received immense attention due to its cost-effective and eco-friendly in nature. However, it has an important impact on bioremediation technology about temperature, pH, oxygen, nutrients and other environmental conditions [5]. Microbial species, substrate properties, chemical structure and so on, also affect bioremediation technology [7]. Additionally, it has limitation of adsorbing sites, which can be overcome by suitable adsorbents. Powdered activated carbon (PAC) is characterized by high surface area and high pore sites for adsorption of pollutants. PAC is also commonly used adsorbent in wastewater treatment worldwide. Other forms of carbons are also used as adsorbent for PAHs (See Table 1) such as carbon fiber, fabrics, petroleum pitch, rayon, etc.

Table 1 List of polycyclic aromatic hydrocarbons (PAHs) removed by carbon materials.

Acenaphthene		
Granular active carbon (GAC) [8] Carbonaceous adsorbents (CAs) [9]	Activated carbon (AC) [10]	Biochar [11-14]
Acenaphthylene		
Granular active carbon (GAC) [8] Carbonaceous adsorbents (CAs) [9]	Activated carbon (AC) [10]	Biochar [12-14]
Anthracene		
Powdered active carbon (PAC) [15] Carbonaceous adsorbents (CAs) [9]	Activated carbon (AC) [10] Corn straw [16] Mesoporous carbon nanoparticles [17]	Biochar [11-14]
Benzo[ghi]perylene		
Carbonaceous adsorbents (CAs) [9]	Activated carbon (AC) [10] Corn straw [16]	Biochar [11-14]
Benz[a]anthracene		
Carbonaceous adsorbents (CAs) [9]	Activated carbon (AC) [10] Corn straw [16]	Biochar [11-14]
Benzo[a]pyrene		
Carbonaceous adsorbents (CAs) [9]	Activated carbon (AC) [10] Corn straw [16] Carbon nano-onion [18]	Biochar [11-14]
Benzo[b]fluoranthene		
Carbonaceous adsorbents (CAs) [9]	Activated carbon (AC) [10] Corn straw [16]	Biochar [11-14]
Benzo[k]fluoranthene		
Carbonaceous adsorbents (CAs) [9]	Activated carbon (AC) [10] Corn straw [16]	Biochar [11-14]
Dibenz[a,h]anthracene		
Carbonaceous adsorbents (CAs) [9] Graphene nanosheets [19]	Activated carbon (AC) [10] Corn straw [16]	Biochar [11-14]

Fluorene		
Granular active carbon (GAC) [8] Carbonaceous adsorbents (CAs) [9]	Activated carbon (AC) [10] Corn straw [16] Carbon nanotubes [20]	Biochar [11-14]
Indeno[1.2.3-cd]pyrene		
Carbonaceous adsorbents (CAs) [9]	Corn straw [16]	Biochar [11-14]
Naphthalene		
Granular active carbon (GAC) [8] Carbonaceous adsorbents (CAs) [9]	Activated carbon (AC) [10,22] Corn straw [16] Carbon nanotubes [20] Coal-based activated carbon (CAC) [23]	Biochar [11-14] Porous carbon materials [24] Graphene oxide [25]
Phenanthrene		
Granular active carbon (GAC) [8] Carbonaceous adsorbents (CAs) [9]	Activated carbon (AC) [10] Corn straw [16] Carbon nano-onion [18] Coal-based activated carbon (CAC) [23]	Biochar [11-14, 21] Graphene oxide [25] Inoculated biochar [26]
Pyrene		
Powdered active carbon (PAC) [1] Carbonaceous adsorbents (CAs) [9]	Activated carbon (AC) [10] Corn straw [16] Coal-based activated carbon (CAC) [23]	Biochar [11-14, 21] Graphene oxide [25] Inoculated biochar [26]

In the present study, immobilized beads were prepared by mixing PAC, sodium alginate (SA) and $CaCl_2$ binder to improve mass transfer in the system for degradation of PAHs. In this study, we have listed all the possible PAHs, from which we have selected pyrene as pollutant for our study. In addition to the immobilized beads, the ratio of PAC, gel, and ratio of crosslinking agent with other immobilized conditions such as pH and temperature are described in this work.

2.   Materials and methods

2.1   Chemicals and culture medium

As described in our previous publication, Luria-Bertani (LB) culture medium (g/L) was prepared by adding 10-g peptone, 5-g yeast extraction, and 10-g NaCl into 1-L deionized (DI) water with maintaining pH 7.2 [5]. Mineral salt medium (MSN) (g/L) was prepared

Advances in Wastewater Treatment I   Materials Research Forum LLC
Materials Research Foundations **91** (2021) 204-218   https://doi.org/10.21741/9781644901144-6

by adding $MgSO_4 \cdot 7H_2O$ 0.25-g, $(NH4)_2SO_4$ 1-g, $NaNO_3$ 2-g, $K_2HPO_4 \cdot 3H_2O$ 10-g, KH_2PO_4 4-g, NaCl 5-g into DI water at pH 7.2. The medium was used for degradation of pyrene by strain. The Pyrene–MSN (g/L) was prepared by introducing 1 mL of the acetone mixed pyrene solution to a conical flask and volatilize acetone by heat, followed by introducing 50 mL MSN medium to a conical flak. In order to make solid beads, 15 to 20-g of agar was added in it to a medium. All the commercially available chemicals were of HPLC grade and used without further purifications.

2.2   Strains and culture

Klebsiella pneumonia (Kp) and *Pseudomonas aeruginosa* (Pa) were screened from sludge of petroleum refinery of Nanjing, China, as described in our previous report [5]. The strain was cultured in conical flask with LB medium at 180 rpm at 30 °C on a shaking table until optical density (OD) OD_{600} (UV-5500 UV spectrophotometer, Shanghai) was obtained 1.0.

2.3   Bacterial strain degradation and PAC adsorption

As described in our previous report, optical density (OD) analysis was used to measure bacterial strain degradation and PAC adsorption at wavelength 600 nm by UV spectrophotometer [5]. In brief, the strain was centrifuged at 180 rpm at 30 °C as described earlier and mixed with 10 mL bacterial suspension, which centrifuged at 2000 rpm for 10 min, discard the supernatant, and use phosphate-buffered saline (PBS) buffer to adjust the substratum into bacterial solution $OD_{600}=2.0$. Then, 0.1 g of PAC was added and adsorbed dynamically in a 25°C shaker. Take the bacterial suspension and settle for 5 min every 1 h, then measure its OD_{600}, and calculate the adsorption amount of the bacterial suspension. Repeat it 3 times and get the average value [27].

2.4   PAC-SA immobilization

The mass fraction of PAC (0.5%, 1%, and 1.5%), SA (2%, 2.5%, and 3%) and $CaCl_2$ (2%, 3%, 4%) were changed to make immobilization beads, and the optimal ratio was obtained by comparing the degradation performance. Briefly, 6% bacterial suspension of PAC was added with 2.5% SA gel and 2% $CaCl_2$ at pH 7.5, with continues stirring for 24h to prevent lump formation. The beads were washed 3-5 times with saline and stored at 4 °C for further use. The effects of operating conditions of immobilization was investigated by changing crosslinking temperature (4,10,15,20,25 and 30°C), crosslinking time(6,12,18,24 and 30h), pH (5,6,7,7.5 and 8) and concentration of bacterial solution (0,2%,4%, 6%,8%, 10%). The mass transfer was determined by measuring luminosity of the solution at 665 nm for 50 beads in 20 mL methylene blue solution for every 10 min of measurement.

2.5   Mechanical strength

Take 4 immobilized microspheres of similar size and shape, place them diagonally, place the glass slide on them, then place a beaker holding water on the glass slide, observe the microsphere until it deforms and cannot recover, and calculate the load bearing capacity. The mechanical strength of the microsphere is indirectly expressed by the mass borne by each microsphere.

3.   Results and discussion

3.1   Bacteria degrade pyrene and PAC adsorption

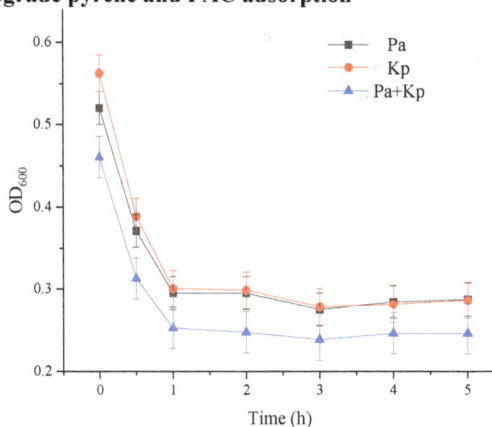

Figure 1: Adsorption curve of activated carbon.

PAC was used to adsorb strains Pa, Kp and bacterial consortium (Pa : Kp =1:1). It was found that OD_{600} value of bacterial suspension of the three groups decreased rapidly in the first 1h (Fig.1). This indicates that the concentration of bacteria is decreasing, that is, a large number of bacteria are adsorbed and deposited on PAC. After 3h, the OD_{600} of the superlative bacteria suspension did not decrease, indicating that the adsorption of PAC was approaching saturation. However, the OD_{600} of bacterial consortium was significantly lower than that of single bacteria ($P < 0.05$), that means the adsorption capacity of PAC to bacterial consortium was higher than that of single bacteria.

Advances in Wastewater Treatment I   Materials Research Forum LLC
Materials Research Foundations **91** (2021) 204-218   https://doi.org/10.21741/9781644901144-6

PAC has a strong adsorption capacity as a non-polar adsorbent. Materials with high adsorption properties can promote the migration of pyrene to the adsorption carrier, and enrich high concentration of microorganisms, which will increase the contact between pyrene and microorganisms. In some studies, *red coccus* immobilized by PAC was in the dominant position in the microbial community, and the removal efficiency of phenol was increased by 5 times [28]. The immobilized hybrid technology has also been applied to remove organic pollutants, which may be the main development direction of immobilized microbial technology in the future. Sekaran et al. found pollutants degradation is the result of the metabolism of bacteria and algae together, which form a mutually beneficial symbiosis in the leather-making wastewater, bacteria provide the carbon source and energy required for algae, and algae in situ generated photosynthetic oxygen for heterotrophic bacteria to mineralize pollutants [29]. Iqbal and Saeed also found that the adsorption immobilization of carrier not only had physical adsorption but also chemical adsorption for pollutants [30]. Microorganisms can absorb organic pollutants adsorbed on carriers and their own surfaces and use them as carbon sources and energy for their metabolism. The effects of immobilized carriers and microorganisms on organic pollutants degradation do not exist alone but in combination. Maleki et al. also found that this degradation process was completed under the synergistic effect of physical adsorption and biodegradation [31].

Table 2 Degradation efficiency of pyrene.

Culturing time	Pyrene removal efficiency (%)			
(day)	CK	Pa	Kp	Pa+Kp
3	4.6	23.6	25.1	28.5
7	6.5	28.7	29.5	35.4
14	7.2	37.5	44.3	48.2

The degradation effect of Pyrene (CK), Pa, Kp and bacterial consortium (Pa:Kp =1:1) after 14d were mentioned in Table 2. It can be seen that the degradation of Kp had higher degradation rates of pyrene than Pa, but bacterial consortium (Pa+Kp) had the best degradation rates of 35.4% and 48.2% on 7 and 14 days, respectively. This indicates that the mixed strains can improve the degradation rate than single microorganism. The obtained results proved that the different enzyme system or metabolic have higher potential to work together to improve degradation.

Advances in Wastewater Treatment I   Materials Research Forum LLC
Materials Research Foundations 91 (2021) 204-218   https://doi.org/10.21741/9781644901144-6

3.2   Embedding and immobilization
3.2.1   Ratio of Gel agent, crosslinking agent and PAC

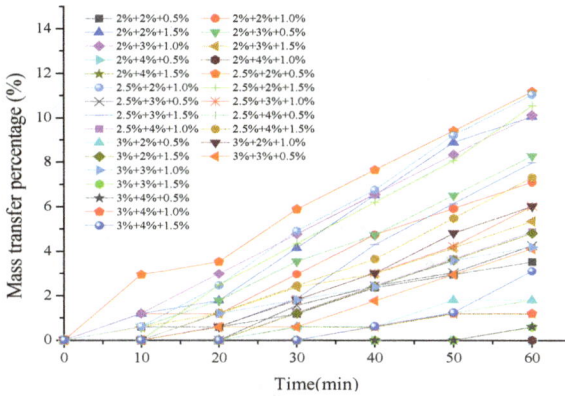

Figure 2: Mass transfer properties of beads with different proportions (SA%+ CaCl₂%+ PAC%).

Fig 2 shows the mass transfer of gel beads with different 27 groups of immobilized beads. Mass transfer of 27 groups of immobilized beads with different proportion was tested within 60 min. The mass transfer of gel beads directly affects enzyme transfer and pyrene degradation of microorganisms. The mass transfer of gel beads was significantly improved after the addition of PAC, and 0.5% PAC- beads was increased by 3% in first 10 minutes (Fig.2). However, the mass transfer was not improved significantly while increasing PAC proportion. The reason was the inefficiency of free bacteria to use low soluble pyrene in soil, and thus, the immobilization strains have more adaptability than free microorganisms. Moreover, presence of PAC can increase porosity and permeability in beads with reducing diffusion resistance of pyrene. However, at higher concentration, mass transfer performance was decreased due to formation of irregular shapes of beads. This may lead to lower mechanical properties, higher swelling in the degradation process and lower strength to beads. This may also be excessive PAC blocking the beads during crosslinking, resulting in incomplete cross-linking of SA and CaCl₂. In common, the embedded ratio was the most important factor for obtaining good mass transfer and properties.

3.2.2 Immobilized condition

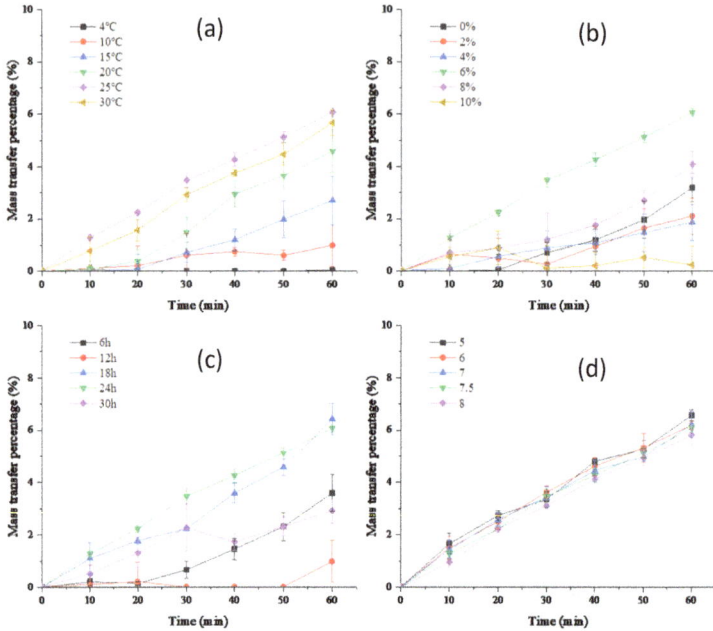

Figure 3: Effect of different experimental conditions (a) temperature, (b) embedded quantity, (c) crosslinking time, (d) crosslinking pH on mass transfer rate.

Factors such as cross-linking temperature, cross-linking time, cross-linking pH and the number of embedded bacteria have important effects on mass transfer of immobilized beads. With the increase of crosslinking temperature, mass transfer of beads improved continuously, and reached the best mass transfer at 25°C (Fig.4a). Temperature affects the density of cross-linking in the process of calcification of beads surface, and then affects the efficiency of material exchange. Fig. 3b shows that the concentration of bacteria had a certain influence on mass transfer, and the embedded quantity of bacteria content 6% increased 6.1% mass transfer in 60 min. It was not conducive to mass transfer and removal efficiency with too high bacterial concentration. The amount of bacteria of beads is an important factor that affects the properties of beads and is directly related to the activity of immobilized bacteria. The larger the amount of bacteria often gets the higher the biological activity, but it was not found in this study. When the amount of bacteria embedded in beads is too large, a large number of bacteria will die because of the lack of space, nutrition and dissolved oxygen. In addition, the high concentration of bacteria will also make the bead pores blocked, which affected the material transfer in and out. Mass transfer of immobilization beads improved with the increase of crosslinking time (Fig.3c). It may be that the immobilization beads becomes denser when the crosslinking time is too long. Moreover, the activity of immobilized bacteria would be affected by the cross-linking effect. Because the SA gel might achieve complete crosslinking with too long crosslinking time, and the mechanical strength of the beads also increased, which leaded the mass transfer of beads decline because of excessive calcification. Mass transfer of beads almost was nothing with pH in immobilization preparation (Fig. 3d). In summary, from the perspective of pyrene removal rate and mass transfer of immobilization beads, the best crosslinking temperature is 25 °C, the best bacterial content is 6%, the best crosslinking time is 24 h, the best crosslinking pH is 7.5.

3.3   Physical properties of immobilized beads.

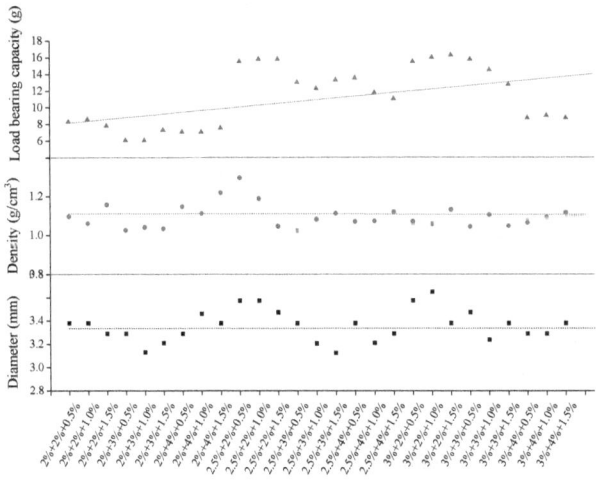

Figure 4: Physical properties of SA-PAC beads (SA%+CaCl$_2$%+ PAC%).

The density, diameter and mechanical strength (load bearing capacity) of beads were checked in the experiment, and the results were shown in Fig. 4. The average diameter of the beads of each group was observed to be 3.4-mm, the average density was found to be 1.1-g /cm^3, and the average mechanical strength was observed to be 11.3-g. It can be seen that the relative fluctuation of diameter and density is small from the trend line, while it has large difference for mechanical strength of immobilization beads. It can significantly improve mechanical strength while proportion 2.5% ~ 3% SA and 2% ~ 3% $CaCl_2$. And the immobilized beads of 2.5%SA + 2%$CaCl_2$ + 0.5% PAC had no damage after 7 days of degradation in the conical bottle in the degradation process for damage rate test. Nedal Massalha et al. found that the smaller the diameter lead the higher the degradation rate of phenanthrene by studying prepared beads of diameter 1, 2, 4 and 6-mm, this because the larger the diameter may be caused by larger resistance of mass transfer [32]. In addition, after adding PAC, direct structure of beads were disturbed, which may responsible to generates pores to obtain nutrients and dissolved oxygen. This behavior will adhere to grow after the microorganism is immobilized by the carrier. The obtained results also support the PAC has a strong adsorption capacity, which can double adsorb the bacteria and substrates, fixing more bacteria, and improving the utilization rate of the bacteria substrates. In general, It is concluded that the distribution pattern is different due to uncontrollable and ununiformed loading of PAC, the possibilities to grow bacterial was higher on the surface, which not affect the other properties but it has ability to enhance mechanical properties. In addition, factors such as carrier concentration and type, crosslinking agent concentration, pH, temperature and others are directly affects the mass transfer.

Conclusion

Bio-retention is the promising and an economical option to remove PAHs from the wastewater system. Our study investigated the contribution of PAC to effectively remove Pyrene pollutant. In this study, 27 different groups of immobilized beads were prepared through different mixing ratio of PAC, SA and $CaCl_2$. In addition, bacterial consortium of Pa, Kp and Pa+Kp were cultured in pyrene -MSN medium for 14 days, bacterial consortium had better degradation effect than single strain. It was 3 h to reach saturation adsorption to strain for PAC, which was beneficial to the growth and reproduction of the strain and promoted the effective contact with pyrene substrate while being added into SA immobilization. The optimum preparation conditions for the SA−PAC immobilization beads were 2.5% SA+ 2% $CaCl_2$+0.5% PAC, adsorption for 3 h, bacterial uptake of 6% (v:v), pH7.5 and 25°C for 24 h. This study confirms the specific ratio of embedded beads were responsible for higher mass transfer and also responsible for higher strength which may further use for degradation of PAHs.

Reference

[1] K. S. Varma, R. J. Tayade, K. J. Shah, P. A. Joshi, A. D. Shukla, V. G. Gandhi, Photocatalytic degradation of pharmaceutical and pesticide compounds (PPCs) using doped TiO2 nanomaterials: A review , Water-Energy Nexus. 3 (2020) 46-61. https://doi.org/10.1016/j.wen.2020.03.008

[2] K. Setty, A. Jiménez, J. Willetts, M. Leifels, J. Bartram, Global water, sanitation and hygiene research priorities and learning challenges under Sustainable Development Goal 6, Development Policy Review. 38.1 (2020) 64-84. https://doi.org/10.1111/dpr.12475

[3] A. Mojiri, J. L.Zhou, A. Ohashi, N. Ozaki, T. Kindaichi, Comprehensive review of polycyclic aromatic hydrocarbons in water sources, their effects and treatments, Science of The Total Environment. 696 (2019) 133971 (1-16). https://doi.org/10.1016/j.scitotenv.2019.133971

[4] C. Grandclement, I. Seyssiecq, A. Piram, P. Wong-Wah-Chung, G. Vanot, N. Tiliacos, N. Roche, P. Doumenq, From the conventional biological wastewater treatment to hybrid processes, the evaluation of organic micropollutant removal: a review, Water Res. 111 (2017) 297-317 https://doi.org/10.1016/j.watres.2017.01.005

[5] Z.You, H. Xu, S. Zhang, H. Kim, P. Chiang, W. Yun, L. Zhang, M. He, Comparison of Petroleum Hydrocarbons Degradation by Klebsiella pneumoniae and Pseudomonas aeruginosa, Applied Sciences, 8. (2018) 2551. https://doi.org/10.3390/app8122551

[6] R. T. Gill, M. J. Harbottle, J. W. Smith, S.F. Thornton, Electrokinetic-enhanced bioremediation of organic contaminants: A review of processes and environmental applications. Chemosphere 107 (2014) 31-42. https://doi.org/10.1016/j.chemosphere.2014.03.019

[7] A. K. Haritash, C. P. Kaushik, Biodegradation aspects of Polycyclic Aromatic Hydrocarbons (PAHs): A review, Journal of Hazardous Materials. 169 (2009) 1-15. https://doi.org/10.1016/j.jhazmat.2009.03.137

[8] D. Eeshwarasinghe, P. Loganathan, M. Kalaruban, D. P. Sounthararajah, J. Kandasamy, S. Vigneswaran, Removing polycyclic aromatic hydrocarbons from water using granular activated carbon: kinetic and equilibrium adsorption studies, Environmental Science and Pollution Research. 25 (2018) 13511-13524. https://doi.org/10.1007/s11356-018-1518-0

[9] F. Li,, J. Chen, X. Hu, F. He, E. Bean, D. C. W. Tsang, Y. S. Ok, and B. Gao, Applications of carbonaceous adsorbents in the remediation of polycyclic aromatic

hydrocarbon-contaminated sediments: A review, Journal of Cleaner Production. 255 (2020) 120263. https://doi.org/10.1016/j.jclepro.2020.120263

[10] Z. Gong, K. Alef, B. Wilke, P. Li, Activated carbon adsorption of PAHs from vegetable oil used in soil remediation: Journal of Hazardous Materials. 143 (2007) 372-378. https://doi.org/10.1016/j.jhazmat.2006.09.037

[11] H. Bao, J. Wang, H. Zhang, J. Li, H. Li, and F. Wu, Effects of biochar and organic substrates on biodegradation of polycyclic aromatic hydrocarbons and microbial community structure in PAHs-contaminated soils: Journal of Hazardous Materials, 385 (2020) 121595. https://doi.org/10.1016/j.jhazmat.2019.121595

[12] B. Tomczyk, A. Siatecka, Y. Gao, Y. S. Ok, A. Bogusz, and P. Oleszczuk, The converting of sewage sludge to biochar as a sustainable tool of PAHs exposure reduction during agricultural utilization of sewage sludges, Journal of Hazardous Materials. 392 (2020) 122416. https://doi.org/10.1016/j.jhazmat.2020.122416

[13] L.Kong, Y. Gao, Q. Zhou, X. Zhao, and Z. Sun, Biochar accelerates PAHs biodegradation in petroleum-polluted soil by biostimulation strategy, Journal of Hazardous Materials. 343 (2018) 276-284. https://doi.org/10.1016/j.jhazmat.2017.09.040

[14] N.Ni, F. Wang, Y. Song, Y. Bian, R. Shi, X. Yang, C. Gu, and X. Jiang, Mechanisms of biochar reducing the bioaccumulation of PAHs in rice from soil: Degradation stimulation vs immobilization, Chemosphere. 196 (2018) 288-296. https://doi.org/10.1016/j.chemosphere.2017.12.192

[15] N. Vainrot, M. S. Eisen, R. Semiat, Membranes in Desalination and Water Treatment: MRS bulletin. 33 (2008) 16-20. https://doi.org/10.1557/mrs2008.9

[16] H. Bao, J. Wang, J. Li, H. Zhang, F. Wu, Effects of corn straw on dissipation of polycyclic aromatic hydrocarbons and potential application of backpropagation artificial neural network prediction model for PAHs bioremediation, Ecotoxicology and Environmental Safety. 186 (2019) 109745. https://doi.org/10.1016/j.ecoenv.2019.109745

[17] M.Kalantari, J. Zhang, Y. Liu, C. Yu, Dendritic mesoporous carbon nanoparticles for ultrahigh and fast adsorption of anthracene, Chemosphere. 215 (2019) 716-724. https://doi.org/10.1016/j.chemosphere.2018.10.071

[18] C. Sakulthaew, S. D. Comfort, C. Chokejaroenrat, X. Li, C. E. Harris, Removing PAHs from urban runoff water by combining ozonation and carbon nano-onions,

Chemosphere. 141 (2015) 265-273.
https://doi.org/10.1016/j.chemosphere.2015.08.002

[19] V. Araujo-Contreras, F. Yepez, O. Castellano, J. Urdaneta, and N. Cubillán,
Interaction of Chrysene, Dibenzo[a,h]anthracene and Dibenzo[a,h]pyrene with
Graphene Models of Different Sizes: Insights from DFT Molecular Electrical
Properties, Polycyclic Aromatic Compounds. 39 (2019) 99-110.
https://doi.org/10.1080/10406638.2016.1267020

[20] A. A.Akinpelu, M. E. Ali, M. R. Johan, R. Saidur, Z. Z. Chowdhury, A. M. Shemsi,
and T. A. Saleh, Effect of the oxidation process on the molecular interaction of
polyaromatic hydrocarbons (PAH) with carbon nanotubes: Adsorption kinetic and
isotherm study, Journal of Molecular Liquids. 289 (2019) 111107.
https://doi.org/10.1016/j.molliq.2019.111107

[21] G. Sigmund, C. Poyntner, G. Piñar, M. Kah, and T. Hofmann, Influence of compost
and biochar on microbial communities and the sorption/degradation of PAHs and
NSO-substituted PAHs in contaminated soils, Journal of Hazardous Materials, 345
(2018) 107-113. https://doi.org/10.1016/j.jhazmat.2017.11.010

[22] S. Bonaglia, E. Broman, B. Brindefalk, E. Hedlund, T. Hjorth, C. Rolff, F. J. A.
Nascimento, K. Udekwu, J. S. Gunnarsson, Activated carbon stimulates microbial
diversity and PAH biodegradation under anaerobic conditions in oil-polluted
sediments, Chemosphere. 248 (2020) 126023.
https://doi.org/10.1016/j.chemosphere.2020.126023

[23] X. Ge, Z. Wu, Z. Wu, Y. Yan, G. Cravotto, B. Ye, Enhanced PAHs adsorption using
iron-modified coal-based activated carbon via microwave radiation, Journal of the
Taiwan Institute of Chemical Engineers. 64 (2016) 235-243.
https://doi.org/10.1016/j.jtice.2016.03.050

[24] J. Yuan, L. Feng, J. Wang, Rapid adsorption of naphthalene from aqueous solution
by naphthylmethyl derived porous carbon materials, Journal of Molecular Liquids. 304
(2020) 112768. https://doi.org/10.1016/j.molliq.2020.112768

[25] X.Yang, H. Cai, M. Bao, J. Yu, J. Lu, and Y. Li, Insight into the highly efficient
degradation of PAHs in water over graphene oxide/Ag3PO4 composites under visible
light irradiation, Chemical Engineering Journal. 334 (2018) 355-376.
https://doi.org/10.1016/j.cej.2017.09.104

[26] B. J. Xiong, Y. C. Zhang, Y. W. Hou, H. P. H. Arp, B. J. Reid, Enhanced
biodegradation of PAHs in historically contaminated soil by M. gilvum inoculated

biochar, Chemosphere. 182 (2017) 316-324.
https://doi.org/10.1016/j.chemosphere.2017.05.020

[27] Y. Y. Sun, X. H. Xu, X. Q. Shi, J. C. Wu, X. H. Li,Biodegradation of Pyrene by Free and Immobilized Cells of Herbaspirillum chlorophenolicum Strain FA1, Water, Air, & Soil Pollution, 227.4 (2016) 120. https://doi.org/10.1007/s11270-016-2824-0

[28] P. Shwu-Ling, H. Yu-Lan, C. Nyuk-Min, S.Ching-Sen, and C. Chei-Hsiang, Continuous degradation of phenol by Rhodococcus sp. immobilized on granular activated carbon and in calcium alginate: Bioresource Technology. 51 (1995) 37-42. https://doi.org/10.1016/0960-8524(94)00078-F

[29] G.Sekaran,, S. Karthikeyan, C. Nagalakshmi, A. B. Mandal, Integrated Bacillus sp. immobilized cell reactor and Synechocystis sp. algal reactor for the treatment of tannery wastewater, Environmental Science and Pollution Research. 20 (2013) 281-291. https://doi.org/10.1007/s11356-012-0891-3

[30] M. Iqbal, and A. Saeed, Biosorption of reactive dye by loofa sponge-immobilized fungal biomass of Phanerochaete chrysosporium, Process Biochemistry. 42 (2007) 1160-1164. https://doi.org/10.1016/j.procbio.2007.05.014

[31] M. Maleki., M. Motamedi, M. Sedighi, S. M. Zamir, and F. Vahabzadeh, Experimental study and kinetic modeling of cometabolic degradation of phenol and p-nitrophenol by loofa-immobilized Ralstonia eutropha, Biotechnology and Bioprocess Engineering. 20 (2015) 124-130. https://doi.org/10.1007/s12257-014-0593-4

[32] N. Massalha, A. Shaviv, and I. Sabbah, Modeling the effect of immobilization of microorganisms on the rate of biodegradation of phenol under inhibitory conditions, Water research. 44 (2010) 5252-5259. https://doi.org/10.1016/j.watres.2010.06.042

Advances in Wastewater Treatment I
Materials Research Foundations **91** (2021) 219-252

Materials Research Forum LLC
https://doi.org/10.21741/9781644901144-7

Chapter 7

New Class of Flocculants and Coagulants

Yongjun Sun*, Shengbao Zhou, Kinjal J. Shah*

College of Urban Construction, Nanjing Tech University, Nanjing, 211816, China

sunyongjun@njtech.edu.cn, kjshah@njtech.edu.cn

Abstract

Coagulation is a kind of efficient water treatment method commonly used in domestic anhydrous and industrial wastewater treatment. Inorganic polymer coagulants (polyvalent metal salts) are widely used because of their low cost and ease of use. However, due to the low flocculation effectiveness and the presence of residual metal concentrations in the treated water, their application is limited. Organic synthetic flocculant has been widely used due to its higher flocculation efficiency at lower dosage. However, it has limitations in applicability due to its molecular structure which is less biodegradable and less disperse in water. Therefore, flocculants based on natural polymers have attracted extensive attention from researchers due to their advantages such as biodegradability and environmental friendliness. This paper summarizes the overview of the development of various types of flocculants that were used for industrial wastewater treatment. In addition, the characteristics and application of flocculant is reviewed with their behavior.

Keywords

Coagulant, Flocculant, Coagulation, Flocculation, Organic/Inorganic Coagulants, Water Treatment

Contents

1.   Introduction

Wastewater from chemical processing industries contains various concentrations of suspended solids, metal ions and other pollutants being smaller in size [1]. Due to the small size of those particles, sedimentation and filtration has become challenge in recent years [2]. Therefore, removing those colloidal particles from the wastewater is a serious challenge for the industry. There are various traditional and advanced technologies been used to remove colloidal particles from wastewater, such as ion exchange, membrane filtration, precipitation, flotation, adsorption, coagulation, flocculation, and electrolysis [3]. Among these methods, coagulation/flocculation is one of the most widely used techniques to remove small particles from wastewater. Moreover, coagulation is a simple and effective wastewater treatment method, and has been widely used to treat various types of wastewater, such as domestic sewage, fine chemical wastewater, papermaking wastewater, oily wastewater, pharmaceutical wastewater, etc. [4, 5]. In the coagulation process, finely dispersed particles, aggregate or coagulate together to form large particles (flocculate) after adding the coagulant and/or flocculant of larger size, which settle and cause clarification of the system [6].

Inorganic polymer coagulant, mainly composed of inorganic metal salts such as iron salts and aluminum salts (such as iron sulfate, ferric chloride, aluminum chloride, etc.) have a long history of use [7]. However, such flocculants are more corrosive and costly. Moreover, under certain conditions, the water purification effect is still not ideal. Compared with traditional agents, inorganic polymer flocculants have better flocculation effect, lower cost, less dosage, and wider pH requirements for wastewater [8]. However, for complex sewage systems, it is difficult to obtain a satisfactory treatment effect using a single flocculant. On the other hand, organic/inorganic composite flocculants can overcome the shortcomings of using a single flocculant to make effective flocculation at lower cost. Composite flocculant has greater advantages than single polymer flocculant, lower cost, better flocculation effect and wider application fields; but still has the disadvantages of producing more sludge and the need of large dosage. In case of the organic flocculants, they are expensive, difficult to degrade, and their residual monomers are toxic in nature. Moreover, its large dosage requirement, and poor stability, attracts the research community to find alternative organic flocculants.

Meanwhile, natural flocculants have received more and more attention due to their low cost, renewable, easily biodegradable and non-toxic nature [10]. So far, various polymer flocculants have been developed or designed to improve the flocculation process of wastewater treatment. Natural flocculants with macromolecule size are widely used in wastewater treatment due to its pH resistance, small dosage, high efficiency, and ease of treatment [11]. However, natural flocculants have moderate efficiency and short shelf life. While, the main problems of synthetic polymer flocculants are non-biodegradability and residual monomer toxicity. Thus, a lot of synthesis and research have been carried out on grafting copolymers to enhance optimal properties [12]. However, finding efficient and economical flocculants has always been a challenge faced by many studies. The main influencing factors of flocculation efficiency include flocculation rate, sedimentation rate, solid sedimentation percentage, turbidity removal rate, and pollutant removal percentage or water recovery rate [13]. These influencing factors actually reflect the size distribution and structure characteristics of flocs during flocculation. Larger, stronger and denser flocs result in improved settling, filtration and clarification performance in wastewater treatment [14].

This article reviews the flocculants studied and applied in wastewater treatment, including inorganic salt coagulants/flocculants, polyacrylamide coagulants, organic-inorganic composite coagulants, and natural modified coagulants. Inorganic salt-type natural modified coagulants are natural raw materials widely developed and utilized in recent years. At the same time, the graft flocculant has been studied recently, and

Advances in Wastewater Treatment I   Materials Research Forum LLC
Materials Research Foundations **91** (2021) 219-252   https://doi.org/10.21741/9781644901144-7

combined with the performance of chemical flocculant and natural flocculant. This article reviews the literature on flocculants and its application in recent years.

2.   Polymer iron coagulant

Inorganic polymer iron-based coagulant is a combination product of the intermediate product of iron ions and other metal ions in the hydrolysis process and the negatively charged sol particles and different anions in the solution. In the actual water treatment process, especially for drinking water, due to the toxicity of aluminum-based coagulants, the research and development of iron-based varieties have become the focus of the development of inorganic polymer coagulants [15]. The polymeric iron coagulant has the advantages of safety and harmlessness, fast flocculation speed, better dehydration of the formed flocs, and high impurity removal rate. It can be used for sewage treatment and drinking water purification. At present, polymeric iron coagulants are mainly divided into traditional polymeric iron coagulants and composite polymeric iron coagulants [16]. Traditional polymeric iron coagulants mainly include polymeric ferric sulfate (PFS), polymeric ferric chloride (PFC), etc. Composite polymeric iron coagulants mainly include polyferric silicate sulfate (PFSS), polyferric silicate chloride (PFSC), polyferric aluminum silicate (PSFA), polyferric aluminum chloride (PAFC), and polyphosphoric iron sulfate (PPFS), Polyferric Chloride Sulfate (PFSC), etc. As the requirements for water quality monitoring become more stringent, a single coagulant can no longer meet the current sewage treatment requirements. Compared with the traditional single iron-containing coagulant, inorganic composite polymeric iron coagulant can achieve good flocculation effect in water and wastewater treatment, while reducing the cost of flocculation, and also expanding the application range of coagulant [16].

Domestic research on iron-based coagulants began in the early 1980s, represented by Tianjin Chemical Industry Research Institute. Since the development of the iron-based coagulant, people have paid more and more attention because of its rapid settling speed and large floc formation. In the research on the flocculation mechanism of the coagulant, since the research of the aluminum-based coagulant is earlier, there is more knowledge about its flocculation mechanism and production process [18]. The iron-based coagulants are rarely reported even in foreign countries. At present, most researchers believe that when iron ions enter water, it forms a polymer with water in a short time. The polymer quickly deprotonates and hydrolyzes into a series of polynuclear iron hydroxide ion monomer forms. Then it is further polymerized into polymer. Since impurities and colloidal ions in water are generally negatively charged, and these inorganic high molecular polymers carry a large amount of positive charge, they quickly neutralize with colloidal ions, destabilizing the colloidal ions. At the same time, this hydrolyzed iron ion has a strong adsorption and bridging effect on particles in water. During the falling process of this polymer, the particulate matter in the water is net-trapped and swept, thereby forming coarse flocs to sink to achieve the purpose of water purification. In addition, in the formed series of multinuclear ions, the ions in the critical state before flocculent precipitation have the best flocculation performance [19]. This may be related to the ability of this critical product to neutralize the colloidal charge, compress the electric double layer, and reduce the colloid potential, which promotes the rapid aggregation and precipitation of colloidal particles and suspended solids [20].

2.1 Traditional polymeric iron coagulant

2.1.1 Polyferric sulfate (PFS)

Polyferric sulfate (PFS) is a polymer iron coagulant that was developed earlier and is also a relatively mature coagulant. The molecular formula can be expressed as $[Fe_2(OH)_n(SO4)_{3-n/2}]m$, which is an intermediate product of ferric sulfate in the hydrolysis-flocculation process, and is a six-coordinate formed by -O- or -OH bridges [21]. The multi-core polymer of iron has an octahedral structure. At present, there are many production processes of polyferric sulfate, such as nitric acid catalytic oxidation method, air oxidation catalytic method, potassium chlorate (sodium chlorate) catalytic oxidation method, one-step method, two-step oxidation method and microbial oxidation method, etc. [21]. The most important factor affecting the quality of PFS is the degree of basicity, namely $[OH^-]/[Fe^{3+}]$. The greater the degree of basicity, the higher the degree of molecular polymerization, the more positive charge the formed hydroxy polymer has, the better the water purification effect. But the basicity is not as large as possible. The basicity of the current mature process is 8% to 15%. Luting [23] used a catalytic oxidation method to prepare a water purifying agent, polyferric sulfate, using ferrous sulfate heptahydrate as a raw material, $NaNO_2$ as a catalyst, and oxygen as an oxidant under acidic conditions. The influence of time, reaction temperature and concentrated sulfuric acid dosage on the preparation of polyferric sulfate, and the effect of the polyferric sulfate prepared by this method on the treatment of municipal sewage was studied. Experiments show that the method has high sensitivity, good selectivity, good retention time and peak area reproducibility, and linear correlation coefficient R>0.999, which satisfies the requirement of sensitive, accurate, and fast analysis methods for water quality detection, and can better serve water quality and ensure the safety of water supply. Liu [24] prepared liquid PFS by air-oxidant combined oxidation method, and studied the effects of the feed ratio, solution pH and reaction temperature on the quality of liquid PFS products. The optimal process for preparing liquid PFS was determined by the feed ratio ($Fe_2SO_4\cdot7H_2O:H_2SO_4$), the solution pH, the reaction temperature, and sodium chlorate. In the printing and dyeing wastewater treatment experiment, when the dosage of liquid PFS was 146 mg•L^{-1}, the COD removal rate was 65.2%, and the turbidity removal rate was 91.4%. It can be seen that the preparation of PFS by different preparation methods has a greater impact on the coagulation performance of PFS.

2.1.2 Polyferric chloride (PFC)

PFC is one of the main varieties of inorganic polymer coagulants. It is an efficient coagulant and has been widely used in water treatment, such as the removal of harmful metals such as arsenic. The coagulation results indicate. Compared with polyferric sulfate, polyferric chloride has the characteristics of fast floc formation, large particle density, good precipitation performance, wide adaptability to water temperature and pH value, and strong COD, BOD and chroma removal ability [25]. There are many methods for preparing PFC. For example, Yang [26] and others use pickling waste liquid as a raw material, and use the catalytic oxidation-polymerization one-step method to prepare. There are also studies on the synthesis of polyferric chloride using nitrogen oxides as catalysts, packed towers as reactors, and pickling waste liquid or iron scraps as iron-containing raw materials. In the preparation process, the acidity of the solution is mainly controlled, part of the iron salt solution is hydrolyzed, the generated hydroxyl group enters into the ferric chloride molecule, and the bridge forms the polyhydroxy core complex—[$Fe_2(OH)_nC_{16-n}$]. However, the stability of PFC is poor, and precipitation usually occurs within a week, which reduces the efficiency of water treatment. Sheng [27] used the pickling wastewater from the iron and steel industry to prepare four polyferric chloride coagulants. The experimental results show that the prepared iron coagulants not only achieve the resource utilization of waste acid, but also become hazardous waste. It is a useful resource, and the iron-based coagulant shows good coagulation effect, and has strong market competitive advantages and application development prospects.

2.2   Composite polymeric iron coagulant

2.2.1 Ferric polysilicate sulfate (PFSS)

Polysilicate ferric sulfate (PFSS) is a new type of water treatment agent developed on the basis of activated silicic acid and polyferric iron. It is a composite inorganic polymer coagulant prepared by introducing metallic iron ions into active silicic acid. The polysilicic acid is used to enhance the adhesion and aggregation ability of the iron polymer, and the gelation time of the polysilicic acid is prolonged by an appropriate amount of iron ions, so the polyferric silicate sulfate has better coagulation performance. Such as turbidity removal, high COD removal efficiency, fast floc sedimentation speed,

Advances in Wastewater Treatment I   Materials Research Forum LLC
Materials Research Foundations **91** (2021) 219-252   https://doi.org/10.21741/9781644901144-7

long stability time and other advantages. The improvement of the properties of coagulant can improve the efficiency of wastewater treatment, save the amount of reagents and reduce secondary pollution, which is conducive to sludge removal and drying treatment. Zhang et al. [29] prepared silica sol by stirring polymerized water glass and acidulant for several hours, and then added ferrous sulfate and sodium chlorate to the silica sol to prepare polymerized ferric silicate by copolymerization. The effects of 4 different acidifiers and different n(Si):(Fe) on its reaction mechanism, stability, surface morphology and coagulation performance were investigated [30]. As a result, it was found that using acetic acid as the acidifying agent can delay the polymerization of silicic acid, and the prepared polymeric silicic acid and polymeric ferric silicate have the best stability. Polysilicate ferric sulfate is a new type of inorganic polymer coagulant synthesized by compounding or copolymerizing iron ions and activated silicic acid using the principle of synergy. Because this coagulant has both electrical neutralization and adsorption bridging effects, with an average relative molecular mass of up to 200,000, it is possible to partially replace organic synthetic polymer coagulants in water treatment to avoid toxicity. Generally, iron ions and aluminum ions are used as coupling metal ions. Compared with polyaluminum silicate, polyiron silicate is a better performance coagulant, because it is basically harmless to the environment, fundamentally Eliminate the problem of residual aluminum in water [31].

2.2.2 Polysilicate ferric chloride (PFSC)

Polyferric chloride is one of the main varieties of inorganic polymer coagulants. The composite coagulant prepared by compounding polysilicic acid and polyferric chloride has both flocculation characteristics, long storage time, convenient use, and greatly reduces corrosion compared with traditional iron salt coagulant. Shi et al. [32] used oligomeric silicic acid as a stabilizer, ferric chloride and sodium bicarbonate as raw materials to prepare polyferric chloride with different degrees of polymerization, and then polymerized with high degree of polymerization Silicic acid is compounded and reacted for 1 to 4 hours to prepare a polymerized ferric chloride coagulant product. Tests show that its flocculation effect is significantly better than ferric chloride. Jian [33] used hydrochloric acid pickling waste liquid as raw materials, and added Pb^{2+} in the precipitated waste liquid of Na_2S, and added sodium silicate for polymerization to prepare polysilicate composite ferric chloride coagulant (PSFC). The influence of Na_2S dosage and reaction temperature on the removal effect of Pb^{2+} was studied, and the reaction kinetics of Pb^{2+} precipitation by Na2S was investigated. The influence of different preparation temperature, n(Fe):n(Si), n(NaOH):n(Fe) on Fe^{3+} polymerization morphology and PSFC flocculation effect was analyzed by single factor experiment method, and the preparation conditions were optimized by orthogonal experiment. The results show that the optimal condition for Na_2S removal of Pb^{2+} is $n(Na_2S):n(Pb^{2+})=3.0$, and the reaction of Na_2S precipitation Pb^{2+} satisfies the first-order reaction kinetic model. The best preparation conditions of coagulant: temperature is 25 °C, $n(Fe):n(Si)=25$, $n(NaOH):n(Fe)=0.5$.

2.2.3 Polyaluminum ferric silicate (PSFA)

As a new inorganic polymer coagulant, polyaluminum ferric silicate (PSAF) has excellent performance in removing turbidity, decoloring and removing organic matter [34]. The key factors to control the performance and stability of PSAF are the Al/Fe/Si molar ratio, silicic acid concentration, alkalinity, curing time, etc. At the same time, the flocculation effect of PSAF is also affected by the water parameters. Polyaluminum ferric silicate introduces two kinds of metal ions into polysilicic acid at the same time. It not only overcomes the PSA's sensitivity to pH, but also has a large amount of residual aluminum, and it also overcomes the shortcomings of PFS effluent with residual color. Common preparation methods of polyaluminum ferric silicate include copolymerization method, compound method and acid solution neutralization method. The difference between the two is whether the cationic solution has polymerized before adding polysilicic acid. The compound method and the copolymerization method affect the morphological distribution of silicon to iron. In the system with the same alkalinity, when the silicon content is large, the iron fraction of the mesoporous state prepared by the compound method is higher than that of the mesoporous state prepared by the copolymerization method [35]. The fraction, while the high iron content in the middle polymer, the low iron content in the high polymer, the polymerization degree is low [36]. The opposite is true when the silicon content is small. Therefore, the experimental method should be determined according to the required mole fraction of iron in the mesomeric state. In actual production, it is usually prepared by the acid leaching neutralization method.

Polysilicic acid is negatively charged and belongs to anionic inorganic polymer material. Introducing Al^{3+}, Fe^{3+}, Al^{3+}, Fe^{3+} into polysilicic acid, the hydrolysate plays a bridging role, so that the charge of polysilicic acid changes from negative to positive, PSAF has electrical neutralization effect on colloidal particles in water, forming a more stable floc body [37]. The addition of polysilicic acid increases the molecular weight of PSAF and improves the adsorption and bridging ability of the coagulant. After the formation of large particles, PSAF's net catching and sweeping effect is more prominent. Especially in water bodies with a high concentration of suspended particles, the net-sweeping action of polymerized silicic acid has a greater effect on the improvement of turbidity removal rate. At present, polyaluminum ferric silicate coagulant can be used to treat printing and dyeing wastewater, domestic sewage, coking wastewater and so on. There are many studies on the influencing factors that affect the flocculation effect, mainly including the Al/Fe/Si molar ratio, the dosage of the coagulant and the pH value of the flocculation treatment. In the polyaluminum ferric silicate coagulant, the Al/Fe/Si molar ratio mainly affects the electric neutralization and adsorption bridging effect [38]. When the ratio is too low, the negatively charged polysilicic acid neutralizes the positive charge of aluminum iron ions, reduces the zeta potential of the electric double layer, reduces the electrical neutralization ability, and affects the flocculation effect. When the ratio is too high, the adsorption bridging effect of polyaluminum ferric silicate coagulant is affected. The charge of the polyaluminum ferric silicate coagulant can neutralize the charge of suspended particles, making the colloid unstable and remove it [39]. When the amount of polyaluminum ferric silicate is too much, the adsorption of suspended particles to the polyaluminum ferric silicate is too large, which may cause the colloid to re-stabilize, or the polyaluminum ferric silicate wraps the suspended particles, and the "colloid protection" effect occurs. Make the flocculation ability worse. When the dosage of coagulant is too small, the adsorption bridging capacity is too weak, it is difficult to connect the rubber bridging bridge, and the flocculation capacity is poor. Therefore, the most suitable dosage of coagulant should not only ensure the rapid flocculation of the colloidal particles, but also make the largest flocculated colloidal particles not easily fall off [40]. In the process of preparing polyaluminum ferric silicate coagulant, pre-hydrolysis experiment was carried out to greatly reduce the hydrolysis reaction in the sewage treatment process. The coagulant can be applied to water bodies with a wider pH range. Adjusting the pH value can change the nature and size of the charged surface of the colloid in water, and affect the hydrolyzability of the coagulant. Generally speaking, the pH value of the water body is within a certain range, and the coagulant has a better treatment effect [41]. Each coagulant has its own suitable pH value range for different water samples. If the pH value is too large, the polynuclear hydroxyl complex ion generated by the hydrolysis of the polyaluminum ferric silicate coagulant decreases the electrical neutralization ability, so the flocculation effect is poor, pH If the value is too low, the protonation effect is obvious, which may cause the zeta potential to change from negative to positive, causing the particles to re-stabilize, thereby affecting the flocculation effect [42].

2.2.4 Polyphosphate ferric sulfate (PPFS)

Polyphosphate ferric sulfate (PPFS) is a new type of inorganic polymer coagulant introduced by PO_4^{3-} synthesis on the basis of polyferric sulfate (PFS). Because PO_4^{3-} can interact with Fe, it can enhance the coordination and coordination ability of PFS and form a multinuclear complex, so it has fast floc formation, large particle density, and high removal rate of turbidity, color, and phosphate [43]. It has the advantages of wide adaptability to water temperature and pH, but it has strong corrosivity and is easy to color water. Tang et al. studied the $FeCl_3$-Na_2HPO_4-$NaHCO_3$ system, and as a result, the surface PO_4^{3-} participated in the hydrolysis and polymerization of Fe^{3+} [44]. Amorphous $Fe(OH)_3$ can adsorb H_2PO_4- and reduce the volume of the solution. These studies indicate that PO_4^{3-} can interact with Fe^{3+} and form a bridge. It can be expected that the introduction of PO_4^{3-} in PFS can increase the degree of polymerization due to the formation of bridge bonds, which is beneficial to improve the flocculation performance of PFS. Based on the polymerization of phosphate on Fe^{3+}, related scholars have also begun to study the preparation and application of polyphosphoric ferric sulfate [45]. There is a method for preparing solid PPFS by reacting solid PFS and $Na_3PO_4 \bullet 12H_2O$, which are ground and mixed uniformly, at 120-180°C. Or use hydrogen peroxide to oxidize ferrous sulfate to obtain PFS, then add Na_3PO_4, and dry to obtain solid PPFS. Studies have shown that PPFS is significantly better than PFS in treating printing and dyeing wastewater, titanium dioxide wastewater, and electroplating wastewater.

3.   Polyaluminum coagulant

Polyaluminum chloride [PAC, $Al_m(OH)_n(H_2O)_x$] is an inorganic polymer coagulant developed in the late 1960s. It is currently the most widely used and sold inorganic coagulant and water treatment agent. The inorganic polymer coagulant represented by PAC is much lower in price than the organic polymer coagulant, and it is easy to store, and the preparation conditions are not as harsh as the organic polymer [47]. Compared with traditional inorganic low-molecular aluminum salts, PAC has more high electric charge, so it has stronger electrical neutralization ability and strong adsorption ability and, shows excellent coagulation effect after being added to water. PAC has the advantages of little influence on the pH of the effluent, less dosage, less sludge, and high turbidity removal [48]. Currently, about 60% of water treatment plants in China use PAC for coagulation treatment. The production process of PAC can be divided into aluminum-containing mineral method, aluminum hydroxide method, aluminum chloride method, and metal aluminum method according to the different raw materials. In addition, there is a method for preparing PAC by using waste molecular sieve, polysilicon residue and red mud as raw materials [49, 50].

As the quality of sewage water becomes more and more complex, a single coagulant can no longer meet the needs of water treatment, the development of cationic composite coagulant has been developed, and a lot of performance, copolymerization mechanism and morphology of cationic composite coagulant have been developed. The representative is iron-aluminum composite, that is, introducing Fe^{3+} into PAC to prepare composite coagulant polyaluminum-iron chloride. PAFC is an inorganic polymer coagulant polymerized by aluminum and iron salts under certain conditions [51]. It has extremely strong electrical neutralization ability, the flocs formed are large and dense, easy to settle, and the raw material source is wide, the cost is low, and the amount of residual aluminum in the effluent is low. It has the characteristics of high basicity of polymerized aluminum salt and strong adaptability to raw water, as well as the advantages of high density of polymerized iron and fast sedimentation of flocs [52]. Therefore, it is widely used in the coagulation treatment of various industrial wastewater, domestic sewage and drinking water.

Although aluminum salt coagulation is widely used in wastewater treatment, aluminum is a substance that is harmful to human health. Once it enters the human body, it will be deposited in certain tissues and cells of the body, which will cause human cells to be unable to absorb various nutrients and trace elements necessary for metabolism, which affects the normal physiological functions of the human body and causes various symptoms of aluminum poisoning [53]. Although the composite inorganic polymer coagulant may enhance the flocculation aggregation effect, improve stability, and prolong the gelation period, the positive charge is significantly reduced. Therefore, in the preparation, it is necessary to take into account the hydrolysis of iron salts and the electrical neutralization capacity of the hydrolysate, and to control their respective optimal polymerization conditions, in order to develop excellent products with large molecular weight, high charge capacity, and strong adaptability [54]. At the same time, the formulation and process should be determined according to the requirements of water quality treatment, and the focus of future research should also be on improving product stability. In addition, the structure of the polysilicate composite coagulant should be characterized with the help of modern analytical instruments, the relationship between the morphological structure and coagulation performance of the polysilicate composite coagulant should be analyzed, and the coagulation mechanism should be further explored [55].

4. Literature references

Polyacrylamide (Polyacrylamide, PAM for short) is a general term for homopolymers and copolymers of Acrylamide (AM for short) and its derivatives. It is a linear water-soluble polymer [56]. It is derived from its molecular structure. Characteristics, PAM has special physical properties and chemical properties, which are not only good in flocculation, surface activity and thickening, but it is also easy to obtain a variety of branched or network structures through grafting or crosslinking [57]. Modified products are widely used in petroleum mining, sewage treatment, papermaking, mining and other industries. According to the different charged properties of polyacrylamide, it can be divided into nonionic polyacrylamide (NPAM), anionic Polyacrylamide (APAM or HPAM) and cationic polymers are collectively called polyacrylamide (PAM). After dissociation in aqueous solution, polyacrylamide can dissociate different charges, according to the different charge properties of the product, it can be divided into four categories: nonionic polyacrylamide (NPAM), anionic polyacrylamide (APAM), cationic polyacrylamide (CPAM) and amphoteric polyacrylamide (AmPAM) [58].

At present, the methods of synthesizing polyacrylamide mainly include aqueous solution polymerization, inverse suspension polymerization, emulsion polymerization, inverse emulsion polymerization, dispersion polymerization, and photo-initiated polymerization. Aqueous solution polymerization refers to the polymerization reaction that takes water as a solvent and induces the monomer solution through an initiator or radiation [59]. This method has the characteristics of simple operation, uniform material mixing, and low production cost, but at the same time, it also has low average molecular weight of the product, difficulty in dispersing the heat of polymerization and prone to burst polymerization, and easy crosslinking of the product during the preparation process, which leads to a decrease in product performance. Disadvantages such as long dissolution time during use. Inverse emulsion polymerization refers to the polymerization of water-soluble monomers, emulsifiers, and initiators dispersed in an oil-soluble medium into an emulsion to obtain a "water-in-oil" (W/O) type emulsion polymer. This method has easy heat transfer, the polymerization rate is fast and easy to control, less gel is generated, the relative molecular weight of the polymer is relatively high, but there are also complex processes, high cost, easy to cause secondary pollution of the environment, and a wide distribution of latex particle size. Disadvantages such as dispersion polymerization means that the monomer, initiator, and stabilizer are all dissolved in the dispersion medium to form a homogeneous system, and the resulting polymer is insoluble in the dispersion medium, and finally precipitates out of the solution. The polymerization method easily dissipates heat [60]. the polymerization process is easy to control and the product has good solubility. At the same time, water-dispersed polymerization greatly reduces the use of organic solvents and surfactants and avoids repeated pollution of the environment. Photo-initiated polymerization is initiated by the system through light excitation in a certain excited state. A crack occurs to generate active radicals to initiate polymerization. Photo-initiated polymerization has the advantages of strong selectivity, rapid initiation at room temperature, easy operation control, and reliable and stable polymerization effect [61].

The flocculation mechanism of polyacrylamide coagulant for impurities in water is mainly divided into three categories: adsorption-electrical neutralization, adsorption bridging, and net-sweeping. The specific mechanism of flocculation depends on the type of coagulant, the size, concentration and nature of the impurity particles in the water, as well as several environmental factors such as pH and ions. It often requires specific analysis. The basis of adsorption-electricity neutralization is still that suspended particles attract differently charged polymers, neutralize the charge carried by the colloidal particles themselves, weaken the electrostatic repulsion between the particles, and thus cause the suspended particles to coagulate spontaneously [62]. Adsorption is also accomplished by electrical or van der Waals forces. Suspended matter in water is generally not prone to aggregation due to the same charge of the particles or has a lower density, so it exists stably in a highly dispersed state. The polymer chain of the polyacrylamide coagulant can attract each other with the suspended particles. If the chain contains groups that are different from the particles in electrical properties or are purely attracted to each other by Edward force, then these particles can interact with coagulant molecules combine to form a structure where a large molecule connects several suspended particles. When polyacrylamide coagulant is put into water, if its molecular chain can be fully stretched, a large number of coagulant molecules can form a fishnet-like structure with each other, and they will come into contact with suspended solids during their natural sinking process, and through mechanical action The particles are rolled into the network structure, or entrapped between the gaps in the molecular chain, so that the particles settle together with the coagulant [63].

Polyacrylamide (PAM) is one of the most important synthetic organic polymer coagulants and one of the components of inorganic-organic composite polymer coagulants. Xu et al. [64] dispersed diatomaceous earth into a solution containing ammonium chloride and acrylamide. The prepared polycationic acrylamide/diatomite composite coagulant can make the treated wastewater the light transmittance exceeds 95%, and the dosage of coagulant is only 7.5mg/L, which is much lower than that of conventional coagulant 60~90mg/L. At the same time, the settling time is less than 5s, similar to conventional coagulants. Li et al. [65] used modified coal gangue and polyacrylamide (PAM) to prepare a new type of coagulant with both turbidity and color removal properties. The flocculation test of oilfield drilling wastewater showed that the turbidity removal rate of the new coagulant is 85.5%, the light transmittance of wastewater after flocculation and sedimentation is 53.6%, and the turbidity removal efficiency is much higher than PAM or PAM/gangue mixture. Xiong et al. [66] used aluminum sulfate, iron sulfate, sodium silicate and cationic polyacrylamide (CPAM) as the main raw materials, and prepared the PSAF-CPAM inorganic-organic composite polymer blend by copolymerization. The results show that the best preparation ratio of PSAF-CPAM is the mass ratio of inorganic coagulant to organic coagulant is 70:1. The removal rate of PSAF-CPAM to the total phosphorus in printing and dyeing wastewater increases with the increase of the dosage of coagulant Gradually increasing, the growth trend slows down after the metal ion concentration of the coagulant dosage reaches 1 mol/L, and the total phosphorus removal rate is above 98%. Electron microscope scanning (SEM) and Fourier infrared analysis (FTIR) were used to characterize and analyze the morphology and structure of PSAF-CPAM coagulant, and it was found that a new polymer formed due to a chemical reaction between PSAF and CPAM. Sun et al. [67] placed acrylamide monomer, ferric chloride hexahydrate and aluminum chloride hexahydrate in $(NH_4)_2S_2O_8$-$NaHSO_3$ solution to synthesize a new type of composite coagulant-poly Acrylamide (PAM)-polyaluminum ferric chloride (PAFC). A composite flocculant synthesized with an initiator mass fraction of 0.5%, a polymerization temperature of 50°C, a monomer mass fraction of 20%, and a polymerization time of 4h was subjected to a flocculation experiment on kaolin-humic acid suspension and synthetic dye wastewater. It shows that the optimal turbidity removal rate is 98.38% when the dosage is 0.6mg/L, and the decolorization efficiency of Congo red and blue-green GL is higher than 93% and 94%, respectively.

5.   Organic/inorganic composite coagulant

At present, with the enhancement of people's awareness of environmental protection and the improvement of water quality requirements, composite polymer coagulants have gradually become the focus of research [68]. At this stage, it is mainly divided into: inorganic-inorganic composite coagulant (including composite coagulants such as aluminum iron, aluminum silicon, etc.), organic composite coagulant, inorganic-organic composite coagulation three categories of agents. Among them, organic-inorganic composite coagulant occupies a dominant position due to its variety, simple process and superior performance [69].

Inorganic-organic composite coagulant is a new type of coagulant developed from the combination of inorganic and organic composite coagulants. It is a combination of inorganic and organic coagulants. It has the dual advantages of inorganic and organic coagulants [70]. While avoiding their respective shortcomings, their performance is well played. The addition of organic polymer coagulant enhances the adsorption and bridging ability of the coagulant, which accelerates the flocculation speed, the flocs are larger, and the settling performance is better. At the same time, it also broadens the scope of use of coagulant, reduces the dosage of coagulant, and reduces the processing cost. Inorganic polymer coagulants can provide a large amount of complex ions in water, and have a strong adsorption effect on particles in water. Due to the low molecular weight, the dosage is generally large. Organic polymer coagulants have a wide range of sources, and the molecular structure contains many functional groups such as carboxyl, amino, or hydroxyl groups [71]. They are easily soluble in water, and have the characteristics of flocculation, thickening, shearing, dispersibility, etc. Strong electric neutrality can accelerate flocculation and precipitation. In order to make up for the shortcomings of more residual inorganic coagulant and smaller molecular weight, inorganic-organic composite polymer coagulant has developed rapidly.

The inorganic components of inorganic-organic composite flocculants are mostly studied in iron salts and aluminum salts. The iron salts are mainly polyferric sulfate (PFS) and polyferric chloride (PFC), and the aluminum salts are polyaluminum chloride (PAC). Organic components mainly include artificial synthesis and natural organic macromolecule polymers. The widely used synthetic organic high molecular polymers include polyacrylamide (PAM), polyvinyl ether, sulfonated polyethylene and so on. Natural organic polymer is the first flocculant discovered and used by humans, but due to its properties of natural substances such as low molecular weight, low charge density, and easily degraded by biodegradation, it is rarely used. Modification of natural organic polymer flocculant can diversify its structure and enhance flocculation performance. In addition, after modification, it is non-toxic or low-toxic, easy to biodegrade, does not produce secondary pollution, and has low price and good quality with application prospects [72].

The composite coagulant of aluminum salt and PAM is one of the most commonly used inorganic-organic composite coagulants. Tang et al. [73] combined PAM with PAC, ferric chloride, alum, and starch to treat the Yangtze River water. The results show that PAM combined with PAC has the best flocculation effect, can effectively remove suspended particles in water, and put the coagulant into use on the Yangtze River passenger ship, which has produced considerable economic benefits. Pinotti et al. [74] using PAC and PAM composite coagulants have strong electrical neutralization ability and adsorption bridging performance, strong adsorption activity, compared with PAC, the removal rate of organic matter can be improved More than 30%. Xue et al. [74] used AlCl3-PAC-PAM composite coagulant to treat papermaking wastewater, and their biochemical oxygen demand (BOD5) and chemical oxygen demand (COD) removal rates were 58.9%, 63.4%, the removal rate of chroma is as high as 75%.

The composite coagulant made by PFS and PAM has the characteristics of low dosage, fast settling and large floc particle size, etc. It has good coagulation and decolorization effect in water treatment. Zeng et al. [76] used PFS-PAM composite coagulant and added sodium diethyldithiocarbamate (DDTC) to treat copper electroplating wastewater, with a copper removal rate exceeding 99.6%. Lee et al. [77] found that the coagulant prepared by the composite of $FeCl_3$ and non-ionic PAM has a better flocculation effect on beverage industrial wastewater and produces less sludge than $FeCl_3$ alone. The amount of sludge produced is reduced by 60%.

Dimethyl diallyl ammonium chloride (DMDAAC) homopolymer (PDMDAAC) and its copolymers are water-soluble cationic coagulants, with high charge density, easy to control molecular weight, low price, and good flocculation effect. The combined use of PDMDAAC and PAC can give full play to the synergistic effect of the two coagulants, improve the flocculation effect, and reduce the treatment cost. Yue et al. [78] used PDMDAAC, a copolymer of PDMDAAC and acrylamide (AM) [P(PDMDAAC-AM)] and PAC to simulate the treatment of kaolin water samples. The results show that when the dosage of PDMDAAC series coagulant is unchanged, the residual turbidity in water gradually decreases with the increase of PAC dosage, and can reach the minimum value when PAC is treated alone, that is, when a small amount of PAC is used It achieves a better flocculation effect and simplifies the dosing and post-treatment process. The composite coagulant of iron series and PDMDAAC has a better treatment effect on low turbid water. Gao et al. [79] found that after PFC and PDMDAAC were combined, the Zeta potential increased, and the effect of pH on the Zeta potential was weakened. In the coagulation treatment of dye wastewater, PFC-PDMDAAC, PFC, and PDMDAAC were added in a step-by-step comparison experiment. The results showed that PFC-PDMDAAC had the highest removal rate of dye. Wei et al. [80] prepared composite coagulants with PFS and PDMDAAC, and found that they have good removal effect on pollutants in water, turbidity of surface water, organic matter, etc.

Modified natural organic polymer flocculants can be divided into starch derivatives, chitin derivatives, natural vegetable gum modification, etc. according to different raw materials. Among them, starch derivatives are widely used due to the advantages of easy availability of raw materials and low prices. However, the treatment effect when used alone is not ideal. Therefore, it is usually used in combination with inorganic flocculants. Ekhtera et al. [81] used composite flocculants synthesized with aluminum-based flocculants and corn starch to simulate the treatment of wastewater. It was found that not only the dosage is reduced, but the turbidity removal effect is good, and the residual amount of aluminum is lower. Ridgway et al. [82] used PAC and cationic modified starch compound flocculant to treat oil refining wastewater, and found that the flocs formed during the coagulation process were large, and significant turbidity removal could be achieved with less dosage and degreasing effect. Chitosan (CTS) is a product of deacetylation of chitin. It can achieve a better treatment effect on wastewater containing metal ions such as chromium, copper, and zinc, but it is expensive and rarely used alone. In order to reduce the treatment cost, PAC was used to prepare composite CTS flocculant. Due to the poor electrical neutralization ability of the CTS molecule, it is also

Advances in Wastewater Treatment I   Materials Research Forum LLC
Materials Research Foundations 91 (2021) 219-252   https://doi.org/10.21741/9781644901144-7

greatly affected by acid and alkali. After the introduction of PAC, the positive charge on the CTS molecular chain and the positive charge on the PAC are superimposed, which significantly enhances the electrical neutralization of the composite flocculant Ability, the flocculation effect has been greatly improved in the application of removing turbidity and heavy metal ions, and the dosage is small, the sedimentation is fast, and the price is low. Pereira et al. [83] used the prepared CTS composite flocculant to treat copper smelting wastewater and found that not only the turbidity removal effect is good, but also some heavy metals can be removed, and the removal rate is greater than 97%.

In the field of water treatment technology, polymerization, compounding, and multifunctionalization have become the current development direction of coagulants. Inorganic-organic polymer composite coagulant has the advantages of perfect preparation technology, large processing capacity, good processing effect, high efficiency and low energy, easy operation and wide application range. It has become the main direction of research and application of composite coagulant. In the future, we should further study the coagulation mechanism of inorganic-organic polymer composite coagulants, optimize the synthesis process, expand the scope of compounding raw materials, give full play to the synergistic effect of inorganic and organic components, and prepare highly targeted and stable quality. High-efficiency, low-cost, green and environment-friendly composite coagulant are best used in water treatment technology.

6.   Natural modified coagulant

Mankind has entered the 21st century, actively promoting green industries, and making "green flocculants" for water treatment has particularly important practical significance. Because the residual monomer-acrylamide contained in PAM is toxic, researchers have been working hard for a long time to reduce the residual monomers in the product, and has already produced advanced products with a monomer content of less than 0.05% in foreign countries [84]. However, to date, it has not been generally recognized as a water purification agent for drinking water plants. Therefore, green natural modified coagulant came into being. Natural modified polymer flocculant has the characteristics and advantages of rich and diversified resources, relatively low raw material prices, non-toxic or low toxicity of raw materials and products, and easy biodegradation after use [84]. Natural polymers have the characteristics of wide molecular weight distribution, many active groups and diversified structures, which are helpful for the development of multifunctional and multi-purpose products with excellent performance. The characteristics and advantages of these natural polymer flocculants make up for the weaknesses and disadvantages of synthetic polymer flocculants [86].

Among the many natural modified coagulants, chitosan and its derivatives have been researched, developed and applied the most. Others include micro-natural modified coagulants (mainly laboratory research), amphoteric flocculants, starch derivatives, cellulose derivatives, lignin derivatives and cationic polyacrylamides [87]. Starch and lignin flocculants have been poorly studied, because of their poor stability. The flocculant containing chitosan and plant gum has been used and developed unprecedentedly in recent years due to its versatility, biocompatibility and stability. The research directions of natural polymer flocculants mainly focus on the following three aspects: First, improve and stabilize the performance of existing natural polymer flocculants, and broaden its application fields. The second is to develop a multifunctional natural polymer flocculant mainly based on cationic groups [88]. The third is to develop and research a new type of flocculant containing organic polymer-inorganic phase compound.

6.1   Chitosan and its derivatives

Most of the pollutants in the water body have a negative charge, so cationic natural polymer flocculants have received more attention. Chitosan is a product of the partial deacetylation of chitin, and chitin is mainly derived from shells [89]. Chitosan contains glucosamine and acetylglucosamine units and is a linear hydrophilic amino polysaccharide molecule. It is difficult to dissolve in water or other organic solvents. In an aqueous solution with a pH lower than 5 or less, chitosan is easily dissolved, and $-NH_2$ in the chitosan molecule is protonated to form $-NH_3^+$, which has the characteristics of a cationic polyelectrolyte and contains a high positive charge density [90]. Moreover, the protonated amino group can form an electrostatic attraction with negatively charged substances such as acid ions, dyes, and organic matter in water. Since chitosan has cationic characteristics, and its molecular chain length and molecular weight are large, these characteristics make chitosan particularly suitable for the treatment of pollutants dissolved in water. As a new type of natural polymer flocculant, chitosan has achieved certain research and development, and has received more and more attention because of its safety, non-toxicity, and easy biodegradation. It is worth noting that the defect of poor solubility of chitosan in neutral and alkaline water environments greatly limits the application of chitosan. Based on the amino and hydroxyl groups on the chitosan molecular chain, chemical groups can be introduced through modification to effectively improve the physical and chemical properties of chitosan, improve water solubility, charge density, relative molecular weight and selectivity, and further expand the flocculant Scope of application [91]. The most widely used modification of chitosan is chemical modification, and there are many chemical modification methods. The main reaction types are acylation, carboxylation, etherification, Schiff reaction, alkylation, and sulfonation, Azide, halogenation, salt formation, integration, hydrolysis, crosslinking, graft copolymerization and other reactions. At present, the chemical modification of chitosan at home and abroad mainly uses the activity of amino and hydroxyl groups on the chitosan molecule to introduce new chemical groups to achieve the purpose of improving the performance of chitosan.

Chitosan natural modified coagulant is used for the treatment of heavy metal wastewater, dye wastewater and other pollutant wastewater. Shuyingjia et al. [92] obtained a new shell by reacting 2,4-bis(dimethylamino)-6-chloro-[1,3,5]-triazine (BDAT) with chitosan Glycan-based flocculant (BDAT-CTS), due to the interaction between the introduced aromatic rings and the existence of charge attraction and coordination, BDAT-CTS has a significant removal rate for the binary pollutants tetracycline TC and Cu(II) improve. Ge et al. [93] used maleated chitosan prepared by microwave irradiation to remove metal ions. Studies have shown that as the degree of maleic anhydride substitution increases, the residual metal ion concentration shows a downward trend. The carboxyl group introduced by maleic anhydride can form more coordination sites to chelate with metal ions, thereby enhancing the removal effect of metal ions. Research by Ma et al. [94] pointed out that although alkylation can effectively improve the solubility of chitosan, the solubility increases with the degree of substitution. However, the substitution of amino groups will affect the reactivity of the modified chitosan flocculant to a certain extent. Huang et al. [95] further proved that succinyl chitosan obtained by introducing succinyl into the amino group of chitosan due to the presence of carboxyl groups can also have cationic dyes in neutral and alkaline environments. Ma et al. [95] used 2-hydroxy-4'-(2-hydroxyethoxy)-2-methylpropanedione as a photoinitiator to promote acrylamide and chitosan through ultraviolet light Sugar nanoparticles synthesized polyacrylamide grafted chitosan nanoparticles (NCS-G-PAM). This reaction shortens the reaction time and lowers the reaction temperature compared to the general free radical reaction. In addition, the grafting of the PAM chain enhances the adsorption and bridging effect, promotes the formation of large shear-resistant flocs between unstable particles, and improves the flocculation effect of the modified chitosan flocculant. Fan et al. [97] compared the effect of magnetic chitosan synthesized by in-situ precipitation and embedding on the removal of Cu(II) and Hg(II) ions. The experimental results show that embedding the magnetic chitosan obtained by the method has a narrower particle size distribution and a larger specific surface area, so it is better than the in-situ precipitation method in removal. Therefore, related studies have begun to improve the preparation conditions of the synthesis and modification of magnetic chitosan to improve its physical and chemical adsorption neutralization ability. Lou et al. [98] used chitosan, acrylamide and fulvic acid for graft copolymerization to synthesize a new flocculant for flocculation experiments on dye wastewater. The study found that at the beginning, the amphoteric copolymer

237

Advances in Wastewater Treatment I   Materials Research Forum LLC
Materials Research Foundations **91** (2021) 219-252   https://doi.org/10.21741/9781644901144-7

neutralized the net charge of the dye and caused the rapid removal of the dye. After obtaining the optimal dose, the resuspension of the suspension occurred due to the electrostatic repulsion between the charged particles. In the case of high pollutant concentration, the bridging flocculation produced by the grafted acrylamide long chain plays a key role.

At present, the mechanism of action of the modified chitosan flocculant is still in its infancy, and the morphology and flocculation kinetics of the floc are lacking in-depth exploration. Factors such as rearrangement of the internal structure of the floc, fractal geometry of the floc and other factors are important factors [99]. The effect of model accuracy has not been fully studied further. Use mathematical models to strengthen the research on the influence of pH, ionic strength, dosage, temperature and other factors on the modified chitosan flocculant, and explore the removal of new environmental pollutants and multi-component pollutants by the synergistic effect of other water treatment methods Potential, broaden the application range of modified chitosan flocculant [100]. Most of the research on modified chitosan flocculants is still in the small-scale laboratory test stage, so it is necessary to develop cheaper and more efficient flocculants to promote future large-scale production and application.

6.2   Starch derivatives

Starch is a green natural renewable resource and an inexhaustible "green organic raw material". Starch has wide sources, large output, and cheap price. It is biodegradable and non-toxic to the environment [101]. Starch and its derivatives were first used in paper, textile, food and other industries, and are an important renewable and biodegradable natural resource. Although starch has certain water solubility, it cannot provide ionic properties as a flocculant and needs to be modified. The methods of starch modification include physical methods and chemical methods, mainly through oxidation, carboxylation, esterification, grafting of other functional groups and etherification of free hydroxyl groups on the starch molecular chain [102]. The modified starch flocculant has the advantages of high efficiency, easy degradation and no secondary pollution. It is an environmentally friendly flocculant, can be used in the water treatment industry, and is a promising polymer flocculant.

The sugar ring monomer in the starch molecule contains many hydroxyl groups, which can be modified by methods such as etherification and grafting to prepare a wide variety of flocculants with excellent performance [103]. Due to the different charge properties of the modifier, the modified starch flocculant can be divided into cationic, anionic and amphoteric types. And because of the different modification methods, its structure can be divided into linear and grafted. Since most of the pollutants in the water show a negative charge state, there are many studies on cationic modified starch flocculants. Hacck et al. [104] prepared 3-chloro-2-lightpropyltrimethylammonium chloride modified cationic linear starch flocculant (St-CTA) by etherification method. Changing the dosage of modifier can change the degree of cation substitution. Through the method of flocculation conditioning, the St-CTA has a good dehydration effect on the suspension of port sediments, and it is used in combination with anionic polyacrylamide to have a better dehydration effect. In the research of Liu et al. [105], this synthesis method was also adopted to prepare a series of St-CTA with different degrees of substitution, and the simultaneous deturbidity and antibacterial performance were investigated. The study found that St-CTA has the dual performance of simultaneously removing turbidity and bacteriostasis. When kaolin and bacteria in water coexist, its flocculation removal has a certain synergistic enhancement effect. The quaternary ammonium salt cation has a strong positive charge property, has the effect of destroying the cell wall of the cell body, and has a significant effect on the cell wall of the Gram-negative bacteria.

Huang et al. [106] prepared methacryloyloxyethyl graft-modified cationic starch flocculant (St-g-PDMC) by grafting method. The copolymer was dehydrated and precipitated by acetone Down, you can improve the purity of the product and reduce toxicity. Compared with starch and polyacrylamide, St-g-PDMC has a good flocculation and purification effect on kaolin suspension. In the low dosage range, the electrical neutralization played a leading role in the flocculation process. In addition, St-g-PDMC can also regulate and dehydrate anaerobic sludge. The conditioned sludge can be easily filtered out, and the dosage is 0.696% of the dry weight of the sludge. St-g-PDMC has very good application prospects for sewage treatment and sludge conditioning. The quaternization of starch can make it have the dual functions of turbidity removal and bacteriostasis, showing broad application prospects.

6.3   Cellulose derivatives

Cellulose is one of the most abundant polysaccharide molecules in the world. In recent years, people have expanded its application in water treatment by modifying it. Sodium carboxymethylcellulose is an anionic water-soluble polyelectrolyte widely used in industrial fields such as food, textile, paper, adhesives, coatings, medicine and cosmetics [107]. The chemical modification of agricultural waste date palm to prepare sodium carboxymethyl cellulose as a flocculant and aluminum sulfate as a coagulant has a good effect on removing the turbidity of drinking water [108]. Compared with the commercial anionic polyacrylamide, the prepared sodium carboxymethylcellulose showed that the sodium carboxymethylcellulose with a degree of substitution of 1.17 and a polymerization degree of 480 had 10% better performance than the commercial anionic polyacrylamide. In the study of Shaabani et al. [109], a sulfonated cellulose was prepared by modifying cotton, using alum as a coagulant, and the sulfonated cellulose as a flocculant, coagulating. After purifying the kaolin suspension and optimizing it by response surface method, the dosage of alum has been reduced. It believes that sulfonated cellulose can be used as a substitute for traditional flocculants. In the study of Suopajarvi et al. [110], an anionic dicarboxylic acid nanocellulose was prepared and used as a flocculant and ferric sulfate as a coagulant to treat municipal wastewater. In the combined use of coagulation and flocculation, the turbidity and COD of the effluent are low, and the dosage of ferric sulfate can be effectively reduced. The anionic dicarboxylic acid nanocellulose exhibits higher stability and wider pH adaptation range in long-term use. In another study by Suopajarvi et al. [111], two types of cationic nanocellulose with different charge densities were prepared by modifying the cellulose of trees. The charge densities were 1.07 and 1.70 mmol/g, respectively. The performance of the cationic nanocellulose on municipal sludge conditioning was investigated and compared with cationic polyacrylamide. The results show that the cationic nanocellulose can effectively condition municipal sludge, and its dosage is close to polyacrylamide. In terms of turbidity removal, the performance of cationic nanocellulose is close to that of cationic polyacrylamide. In terms of COD removal, the performance of cationic nanocellulose is higher than that of cationic polyacrylamide [112].

Conclusion

The potential applications of inorganic polymer coagulants, synthetic polymer coagulants, organic/inorganic composite coagulants and natural modified coagulants in wastewater treatment have been summarized in depth. In some studies, it has a significant removal effect on water quality parameters such as suspended solids, turbidity, COD and color, and the removal rate is generally more than 90%. As far as traditional inorganic coagulants are concerned, due to the high complexity of the flocculation process and the variety of polyelectrolytes available, there are still few industrial practices for optimization of flocculation. One way to optimize the flocculation process is to select or control the molecular weight and charge density range of the polymer. Different molecular weights and charge densities produce different flocculation mechanisms. Research is needed to investigate how molecular weight and charge density distribution affect flocculation performance in order to provide better flocculant choices for specific industrial applications. Optimizing these factors can significantly improve processing efficiency and reduce chemical costs.

Since the use of conventional flocculants is closely related to environmental pollution and health hazards, it is necessary to synthesize environmentally friendly, economically

Advances in Wastewater Treatment I   Materials Research Forum LLC
Materials Research Foundations **91** (2021) 219-252   https://doi.org/10.21741/9781644901144-7

viable flocculants with high flocculation efficiency. Some of the developed natural modified coagulants have good flocculation performance, and have excellent removal effects on suspended solids, turbidity, COD and color in various wastewaters. According to the actual industrial production and application, the cost-benefit analysis and optimization of natural modified coagulants are needed to produce a standard production protocol and maximize the flocculation efficiency. It is necessary to judge the economics of its actual use. In addition, the work carried out on an industrial scale is very limited, focusing mainly on laboratory testing. Finally, choosing a high-efficiency flocculant that can remove or reduce almost all pollutants in wastewater is the key to the success of the flocculation process. From two aspects of performance and cost of flocculant, it is a promising flocculant material to produce environmentally friendly flocculant with high removal rate and high concentration flocculant by simple, economical and feasible process.

References

[1] H. Zheng, Y. Sun, C. Zhu, J. Guo, C. Zhao, Y. Liao, and Q. Guan, UV-initiated polymerization of hydrophobically associating cationic flocculants: Synthesis, characterization, and dewatering properties. Chem. Eng. J. 234 (2013) 318-326. https://doi.org/10.1016/j.cej.2013.08.098

[2] Y.J. Sun, H.L. Zheng, M.Z. Tan, J.Y. Ma, W. Fan, and Y. Liao, Synthesis and Application of Hydrophobically Associating Cationic Polyacrylamide. Asian Journal of Chemistry 26 (2014) 3769-3773. https://doi.org/10.14233/ajchem.2014.15948

[3] H. Zheng, Y. Sun, J. Guo, F. Li, W. Fan, Y. Liao, and Q. Guan, Characterization and Evaluation of Dewatering Properties of PADB, a Highly Efficient Cationic Flocculant. Ind. Eng. Chem. Res. 53 (2014) 2572-2582. https://doi.org/10.1021/ie403635y

[4] S. Li, Z. Wu, H. Tang, and J. Yang, Selective adsorption of protein on micropatterned flexible poly(ethylene terephthalate) surfaces modified by vacuum ultraviolet lithography. Appl. Surf. Sci. 258 (2012) 4222-4227. https://doi.org/10.1016/j.apsusc.2011.12.027

[5] C.S. Lee, J. Robinson, and M.F. Chong, A review on application of flocculants in wastewater treatment. Process Saf. Environ. 92 (2014) 489-508. https://doi.org/10.1016/j.psep.2014.04.010

[6] J. Wang, J. Yang, H.W. Zhang, W.S. Guo, and H.H. Ngo, Feasibility study on magnetic enhanced flocculation for mitigating membrane fouling. J. Ind. Eng. Chem. 26 (2015) 37-45. https://doi.org/10.1016/j.jiec.2014.11.038

[7] A. Nawaz, Z. Ahmed, A. Shahbaz, Z. Khan, and M. Javed, Coagulation-flocculation for lignin removal from wastewater - a review. Water Sci. Technol. 69 (2014) 1589-1597. https://doi.org/10.2166/wst.2013.768

[8] A.K. Verma, R.R. Dash, and P. Bhunia, A review on chemical coagulation/flocculation technologies for removal of colour from textile wastewaters. J. Environ. Manage. 93 (2012) 154-168. https://doi.org/10.1016/j.jenvman.2011.09.012

[9] R. Yang, H.J. Li, M. Huang, H. Yang, and A.M. Li, A review on chitosan-based flocculants and their applications in water treatment. Water Res. 95 (2016) 59-89. https://doi.org/10.1016/j.watres.2016.02.068

[10] A.J. Harford, A.C. Hogan, D.R. Jones, and R.A. van Dam, Ecotoxicological assessment of a polyelectrolyte flocculant. Water Res. 45 (2011) 6393-6402. https://doi.org/10.1016/j.watres.2011.09.032

[11] A.Y. Zahrim, C. Tizaoui, and N. Hilal, Coagulation with polymers for nanofiltration pre-treatment of highly concentrated dyes: A review. Desalination 266 (2011) 1-16. https://doi.org/10.1016/j.desal.2010.08.012

[12] D.H. Bache, and R. Gregory, Flocs and separation processes in drinking water treatment: a review. J. Water Supply Res. T. 59 (2010) 16-30. https://doi.org/10.2166/aqua.2010.028

[13] F. Renault, B. Sancey, P.M. Badot, and G. Crini, Chitosan for coagulation/flocculation processes - An eco-friendly approach. Eur. Polym. J. 45 (2009) 1337-1348. https://doi.org/10.1016/j.eurpolymj.2008.12.027

[14] F. Renault, B. Sancey, P.M. Badot, and G. Crini, Chitosan for coagulation/flocculation processes - An eco-friendly approach. Eur. Polym. J. 45 (2009) 1337-1348. https://doi.org/10.1016/j.eurpolymj.2008.12.027

[15] Y. Sun, C. Zhu, H. Zheng, W. Sun, Y. Xu, X. Xiao, Z. You, and C. Liu, Characterization and coagulation behavior of polymeric aluminum ferric silicate for high-concentration oily wastewater treatment. Chemical Engineering Research and Design 119 (2017) 23-32. https://doi.org/10.1016/j.cherd.2017.01.009

[16] Z.Y. You, H.Y. Xu, Y.J. Sun, S.J. Zhang, and L. Zhang, Effective treatment of emulsified oil wastewater by the coagulation- flotation process. RSC Adv. 8 (2018) 40639-40646. https://doi.org/10.1039/C8RA06565A

[17] Y. Sun, A. Chen, W. Sun, K.J. Shah, H. Zheng, and C. Zhu, Removal of Cu and Cr ions from aqueous solutions by a chitosan-based flocculant. Desalin. Water Treat. 148 (2019) 259-269. https://doi.org/10.5004/dwt.2019.23953

[18] J. Ma, K. Fu, X. Fu, Q. Guan, L. Ding, J. Shi, G. Zhu, X. Zhang, S. Zhang, and L. Jiang, Flocculation properties and kinetic investigation of polyacrylamide with different cationic monomer content for high turbid water purification. Sep. Purif. Technol. 182 (2017) 134-143. https://doi.org/10.1016/j.seppur.2017.03.048

[19] X. Zhang, J. Ma, K. Fu, X. Fu, L. Ding, Q. Guan, J. Shi, and L. Jiang, Research on Synthesis of Nano Chitosan modified Polyacrylamide through Low-pressure Ultraviolet Initiation. J. Polym. Mater. 34 (2017) 129-143.

[20] J. Ma, K. Fu, J. Shi, Y. Sun, X. Zhang, and L. Ding, Ultraviolet-assisted synthesis of polyacrylamide-grafted chitosan nanoparticles and flocculation performance. Carbohyd. Polym. 151 (2016) 565-575. https://doi.org/10.1016/j.carbpol.2016.06.002

[21] H. Zheng, G. Zhu, S. Jiang, T. Tshukudu, X. Xiang, P. Zhang, and Q. He, Investigations of coagulation-flocculation process by performance optimization, model prediction and fractal structure of flocs. Desalination 269 (2011) 148-156. https://doi.org/10.1016/j.desal.2010.10.054

[22] G. Zhu, H. Zheng, Z. Zhang, T. Tshukudu, P. Zhang, and X. Xiang, Characterization and coagulation-flocculation behavior of polymeric aluminum ferric sulfate (PAFS). Chem. Eng. J. 178 (2011) 50-59. https://doi.org/10.1016/j.cej.2011.10.008

[23] P. Luting, and W. Jinfeng, Research and progress of the preparation technologies of polyferric sulphate [J]. Industrial Water Treatment 9 (2009) 147-150.

[24] C. Liu, Y. He, F. Li, and H. Wang, Preparation of poly ferric sulfate and the application in micro-polluted raw water treatment. Journal of the Chinese Advanced Materials Society 1 (2013) 210-218. https://doi.org/10.1080/22243682.2013.835120

[25] G. Zhu, H. Zheng, W. Chen, W. Fan, P. Zhang, and T. Tshukudu, Preparation of a composite coagulant: Polymeric aluminum ferric sulfate (PAFS) for wastewater treatment. Desalination 285 (2012) 315-323. https://doi.org/10.1016/j.desal.2011.10.019

[26] Z. Yang, X. Lu, B. Gao, Y. Wang, Q. Yue, and T. Chen, Fabrication and characterization of poly (ferric chloride)-polyamine flocculant and its application to the decolorization of reactive dyes. J. Mater. Sci. 49 (2014) 4962-4972. https://doi.org/10.1007/s10853-014-8197-0

[27] X. Sheng, Study on Poly Ferric Chloride from Steel Pickling Acid Waste. Shandong Chemical Industry 6 (2012) 10.

[28] T. Tshukudu, H.L. Zheng, X.B. Hua, J. Yang, M.Z. Tan, J.Y. Ma, Y.J. Sun, and G.C. Zhu, Response surface methodology approach to optimize coagulation-flocculation process using composite coagulants. Korean J. Chem. Eng. 30 (2013) 649-657. https://doi.org/10.1007/s11814-012-0169-y

[29] W. Zhang, L. Yao, J. Ma, and D. Li, Study on Preparation and Flocculation Properties of Inorganic Polymer Flocculant Polyferric Silicate Sulfate (PFSS), 2010 4th International Conference on Bioinformatics and Biomedical Engineering, IEEE, Chengdu, 2010, pp. 1-4. https://doi.org/10.1109/ICBBE.2010.5517958

[30] T. Tshukudu, H. Zheng, and J. Yang, Optimization of Coagulation with PFS-PDADMAC Composite Coagulants Using the Response Surface Methodology Experimental Design Technique. Water Environ. Res. 85 (2013) 456-465. https://doi.org/10.2175/106143012X13560205144515

[31] T. Tshukudu, H. Zheng, X. Hua, J. Yang, M. Tan, J. Ma, Y. Sun, and G. Zhu, Response surface methodology approach to optimize coagulation-flocculation process using composite coagulants. Korean J. Chem. Eng. 30 (2013) 649-657. https://doi.org/10.1007/s11814-012-0169-y

[32] J. Shi, Q. Liu, and D.S. Wang, Studies on the realationship between the speciation distribution and Si/Fe ratio of poly-silica-ferric-chloride, Advanced Materials Research, Trans Tech Publ, Wu Han, 2011, pp. 1339-1342. https://doi.org/10.4028/www.scientific.net/AMR.156-157.1339

[33] C. Jian, Research Progress of Inorganic Polymer Flocculants. Journal of Langfang Teachers University (Natural Science Edition) 16 (2016) 70-72.

[34] H.L. Zheng, J.Y. Ma, C.J. Zhu, Z. Zhang, L.W. Liu, Y.J. Sun, and X.M. Tang, Synthesis of anion polyacrylamide under UV initiation and its application in removing dioctyl phthalate from water through flocculation process. Sep. Purif. Technol. 123 (2014) 35-44. https://doi.org/10.1016/j.seppur.2013.12.018

[35] L. Zhang, Y. Zeng, and Z. Cheng, Removal of heavy metal ions using chitosan and modified chitosan: A review. J. Mol. Liq. 214 (2016) 175-191. https://doi.org/10.1016/j.molliq.2015.12.013

[36] R. Wang, K. Sun, J. Wang, Y. He, P. Song, and Y. Xiong, Preparation and Application of Natural Polymer/Hydroxyapatite Composite. Prog. Chem. 28 (2016) 885-895.

Materials Research Forum LLC
https://doi.org/10.21741/9781644901144-7

[37] C. Feng, X. Ge, D. Wang, and H. Tang, Effect of aging condition on species transformation in polymeric Al salt coagulants. Colloid. Surface. A. 379 (2011) 62-69. https://doi.org/10.1016/j.colsurfa.2010.11.046

[38] C. Feng, H. Tang, and D. Wang, Differentiation of hydroxyl-aluminum species at lower OH/Al ratios by combination of 27Al NMR and Ferron assay improved with kinetic resolution. Colloid. Surface. A. 305 (2007) 76-82. https://doi.org/10.1016/j.colsurfa.2007.04.043

[39] B. Shi, G. Li, D. Wang, C. Feng, and H. Tang, Removal of direct dyes by coagulation: The performance of preformed polymeric aluminum species. J. Hazard. Mater. 143 (2007) 567-574. https://doi.org/10.1016/j.jhazmat.2006.09.076

[40] B. Shi, Q. Wei, D. Wang, Z. Zhu, and H. Tang, Coagulation of humic acid: The performance of preformed and non-preformed Al species. Colloid. Surface. A. 296 (2007) 141-148. https://doi.org/10.1016/j.colsurfa.2006.09.037

[41] B. Shi, G. Li, D. Wang, and H. Tang, Separation of Al-13 from polyaluminum chloride by sulfate precipitation and nitrate metathesis. Sep. Purif. Technol. 54 (2007) 88-95. https://doi.org/10.1016/j.seppur.2006.08.011

[42] C. Ye, D. Wang, B. Shi, J. Yu, J. Qu, M. Edwards, and H. Tang, Alkalinity effect of coagulation with polyaluminum chlorides: Role of electrostatic patch. Colloid. Surface. A. 294 (2007) 163-173. https://doi.org/10.1016/j.colsurfa.2006.08.005

[43] X. Wu, X. Ge, D. Wang, and H. Tang, Distinct coagulation mechanism and model between alum and high Al-13-PACl. Colloid. Surface. A. 305 (2007) 89-96. https://doi.org/10.1016/j.colsurfa.2007.04.046

[44] H. Tang, F. Xiao, and D. Wang, Speciation, stability, and coagulation mechanisms of hydroxyl aluminum clusters formed by PACl and alum: A critical review. Adv. Colloid Interfac. 226 (2015) 78-85. https://doi.org/10.1016/j.cis.2015.09.002

[45] X. Wu, C. Ye, D. Wang, X. Ge, and H. Tang, Effect of speciation transformation on the coagulation behavior of Al-13 and Al-13 aggregates. Water Sci. Technol. 59 (2009) 815-822. https://doi.org/10.2166/wst.2009.064

[46] H. Liu, D. Wang, M. Wang, H. Tang, and M. Yang, Effect of pre-ozonation on coagulation with IPF-PACls: Role of coagulant speciation. Colloid. Surface. A. 294 (2007) 111-116. https://doi.org/10.1016/j.colsurfa.2006.08.008

[47] Z. Bi, C. Feng, D. Wang, X. Ge, and H. Tang, Transformation of planar Mogel Al-13 to epsilon Keggin Al-13 in dissolution process. Colloid. Surface. A. 407 (2012) 91-98. https://doi.org/10.1016/j.colsurfa.2012.05.013

[48] J. Cao, Z. Wu, S. Li, H. Tang, and Q. Mei, Site-selective adsorption of protein induced by a metal pattern on a poly(ethylene terephthalate) surface. Colloid. Surface. B. 111 (2013) 418-422. https://doi.org/10.1016/j.colsurfb.2013.06.023

[49] C. Feng, Q. Wei, S. Wang, B. Shi, and H. Tang, Speciation of hydroxyl-Al polymers formed through simultaneous hydrolysis of aluminum salts and urea. Colloid. Surface. A. 303 (2007) 241-248. https://doi.org/10.1016/j.colsurfa.2007.04.005

[50] T. Li, Z. Zhu, D. Wang, C. Yao, and H. Tang, The strength and fractal dimension characteristics of alum-kaolin flocs. Int. J. Miner. Process. 82 (2007) 23-29. https://doi.org/10.1016/j.minpro.2006.09.012

[51] Y. Wang, H. Zhou, F. Yu, B. Shi, and H. Tang, Fractal adsorption characteristics of complex molecules on particles - A case study of dyes onto granular activated carbon (GAC). Colloid. Surface. A. 299 (2007) 224-231. https://doi.org/10.1016/j.colsurfa.2006.11.044

[52] T. Li, D. Wang, B. Zhang, H. Liu, and H. Tang, Morphological characterization of suspended particles under wind-induced disturbance in Taihu Lake, China. Environ. Monit. Assess. 127 (2007) 79-86. https://doi.org/10.1007/s10661-006-9261-2

[53] X. Wu, D. Wang, X. Ge, and H. Tang, Coagulation of silica microspheres with hydrolyzed Al(III)-Significance of Al-13 and Al-13 aggregates. Colloid. Surface. A. 330 (2008) 72-79. https://doi.org/10.1016/j.colsurfa.2008.07.034

[54] D. Wang, J. Gregory, and H. Tang, Mechanistic difference of coagulation of kaolin between PACl and cationic polyelectrolytes: A comparative study on zone 2 coagulation. Dry. Technol. 26 (2008) 1060-1067. https://doi.org/10.1080/07373930802179327

[55] C. Feng, S. Zhao, Z. Bi, D. Wang, and H. Tang, Speciation of prehydrolyzed Al salt coagulants with electrospray ionization time-of-flight mass spectrometry and Al-27 NMR spectroscopy. Colloid. Surface. A. 392 (2011) 95-102. https://doi.org/10.1016/j.colsurfa.2011.09.039

[56] Y. Sun, M. Ren, W. Sun, X. Xiao, Y. Xu, H. Zheng, H. Wu, Z. Liu, and H. Zhu, Plasma-induced synthesis of chitosan-g-polyacrylamide and its flocculation performance for algae removal. Environ. Technol. 40 (2017) 954-968. https://doi.org/10.1080/09593330.2017.1414312

[57] Y. Sun, J. Liu, W. Sun, H. Zheng, and K.J. Shah, An alternative strategy for enhanced algae removal by cationic chitosan-based flocculants. Desalin. Water Treat. 167 (2019) 13-26. https://doi.org/10.5004/dwt.2019.24636

[58] X. Xiao, Y. Sun, W. Sun, H. Shen, H. Zheng, Y. Xu, J. Zhao, H. Wu, and C. Liu, Advanced treatment of actual textile dye wastewater by Fenton-flocculation process. The Canadian Journal of Chemical Engineering 95 (2017) 1245-1252. https://doi.org/10.1002/cjce.22752

[59] L. Feng, J. Liu, C. Xu, W. Lu, D. Li, C. Zhao, B. Liu, X. Li, S. Khan, H. Zheng, and Y. Sun, Better understanding the polymerization kinetics of ultrasonic-template method and new insight on sludge floc characteristics research. Sci. Total Environ. 689 (2019) 546-556. https://doi.org/10.1016/j.scitotenv.2019.06.475

[60] S. Zhang, H. Zheng, X. Tang, Y. Sun, Y. Wu, X. Zheng, and Q. Sun, Evaluation a self-assembled anionic polyacrylamide flocculant for the treatment of hematite wastewater: Role of microblock structure. J. Taiwan Inst. Chem. E. 95 (2019) 11-20. https://doi.org/10.1016/j.jtice.2018.09.030

[61] L. Feng, J. Liu, C. Xu, W. Lu, D. Li, C. Zhao, B. Liu, X. Li, S. Khan, H. Zheng, and Y. Sun, Better understanding the polymerization kinetics of ultrasonic-template method and new insight on sludge floc characteristics research. The Science of the total environment 689 (2019) 546-556. https://doi.org/10.1016/j.scitotenv.2019.06.475

[62] Y. Sun, K.J. Shah, W. Sun, and H. Zheng, Performance evaluation of chitosan-based flocculants with good pH resistance and high heavy metals removal capacity. Sep. Purif. Technol. 215 (2019) 208-216. https://doi.org/10.1016/j.seppur.2019.01.017

[63] Y. Sun, W. Sun, K.J. Shah, P. Chiang, and H. Zheng, Characterization and flocculation evaluation of a novel carboxylated chitosan modified flocculant by UV initiated polymerization. Carbohyd. Polym. 208 (2019) 213-220. https://doi.org/10.1016/j.carbpol.2018.12.064

[64] K. Xu, Y. Liu, Y. Wang, Y. Tan, X. Liang, C. Lu, H. Wang, X. Liu, and P. Wang, A novel poly (acrylic acid-co-acrylamide)/diatomite composite flocculant with outstanding flocculation performance. Water Sci. Technol. 72 (2015) 889-895. https://doi.org/10.2166/wst.2015.290

[65] J. Li, J. Li, X. Liu, Z. Du, and F. Cheng, Effect of silicon content on preparation and coagulation performance of poly-silicic-metal coagulants derived from coal gangue for coking wastewater treatment. Sep. Purif. Technol. 202 (2018) 149-156. https://doi.org/10.1016/j.seppur.2018.03.055

[66] X. Xiong, L. Wei, X.U. Xia, and D. Yan, Synthesis of Polymeric Hybrid Flocculant PSAF-CPAM and its Phosphorus Removal in Printing and Dyeing Wastewater. Industrial Safety and Environmental Protection 6 (2018) 21.

[67] Y. Sun, H. Zheng, M. Tan, Y. Wang, X. Tang, L.I. Feng, and X. Xiang, Synthesis and characterization of composite flocculant PAFS-CPAM for the treatment of textile dye wastewater. J. Appl. Polym. Sci. 131 (2014) 156-161. https://doi.org/10.1002/app.40062

[68] H. Salehizadeh, N. Yan, and R. Farnood, Recent advances in polysaccharide bio-based flocculants. Biotechnol. Adv. 36 (2018) 92-119. https://doi.org/10.1016/j.biotechadv.2017.10.002

[69] M. Sillanpaa, M.C. Ncibi, A. Matilainen, and M. Vepsalainen, Removal of natural organic matter in drinking water treatment by coagulation: A comprehensive review. Chemosphere 190 (2018) 54-71. https://doi.org/10.1016/j.chemosphere.2017.09.113

[70] J. Desbrières, and E. Guibal, Chitosan for wastewater treatment. Polym. Int. 67 (2018) 7-14. https://doi.org/10.1002/pi.5464

[71] L.Y. Chai, Q.Z. Li, Q.W. Wang, and X. Yan, Solid-liquid separation: an emerging issue in heavy metal wastewater treatment. Environ. Sci. Pollut. R. 25 (2018) 17250-17267. https://doi.org/10.1007/s11356-018-2135-7

[72] P. Kanmani, J. Aravind, M. Kamaraj, P. Sureshbabu, and S. Karthikeyan, Environmental applications of chitosan and cellulosic biopolymers: A comprehensive outlook. Bioresource Technol. 242 (2017) 295-303. https://doi.org/10.1016/j.biortech.2017.03.119

[73] Q. Tang, J. Wu, J. Lin, Q. Li, and S. Fan, Two-step synthesis of polyacrylamide/polyacrylate interpenetrating network hydrogels and its swelling/deswelling properties. J. Mater. Sci. 43 (2008) 5884-5890. https://doi.org/10.1007/s10853-008-2857-x

[74] A. Pinotti, and N. Zaritzky, Effect of aluminum sulfate and cationic polyelectrolytes on the destabilization of emulsified wastes. Waste Manage. 21 (2001) 535-542. https://doi.org/10.1016/S0956-053X(00)00110-0

[75] M.X. Xue, B.Y. Gao, X. Xu, and W. Song, Polyamidine as a New-Style Coagulant Aid for Dye Wastewater Treatment and its Floc Characteristics, Materials Science Forum, Trans Tech Publ, Singapore, 2018, pp. 930-940. https://doi.org/10.4028/www.scientific.net/MSF.913.930

[76] K. Zeng, W. Qin, F. Jiao, M. He, and L. Kong, Treatment of mine drainage generated by lead-zinc concentration plant. J. Cent. South Univ. 21 (2014) 1453-1460. https://doi.org/10.1007/s11771-014-2085-2

[77] K.E. Lee, T.T. Teng, N. Morad, B.T. Poh, and M. Mahalingam, Flocculation activity of novel ferric chloride-polyacrylamide (FeCl3-PAM) hybrid polymer. Desalination 266 (2011) 108-113. https://doi.org/10.1016/j.desal.2010.08.009

[78] Q. YUE, Y. LI, B. GAO, Z. YANG, and X. ZOU, Study on the inverse emulsion polymerization of PDMDAAC-AM [J]. Journal of Shandong University (Natural Science) 6 (2004) 120-125.

[79] B. Gao, Y. Wang, Q. Yue, J. Wei, and Q. Li, Color removal from simulated dye water and actual textile wastewater using a composite coagulant prepared by ployferric chloride and polydimethyldiallylammonium chloride. Sep. Purif. Technol. 54 (2007) 157-163. https://doi.org/10.1016/j.seppur.2006.08.026

[80] J.C. Wei, B.Y. Gao, Q.Y. Yue, Y. Wang, and L. Lu, Performance and mechanism of polyferric-quaternary ammonium salt composite flocculants in treating high organic matter and high alkalinity surface water. J. Hazard. Mater. 165 (2009) 789-795. https://doi.org/10.1016/j.jhazmat.2008.10.069

[81] M.H. Ekhtera, P.R. Charani, O. Ramezani, and M. Azadfallah, EFFECTS OF POLY-ALUMINUM CHLORIDE, STARCH, ALUM AND ROSIN ON THE ROSIN SIZING, STRENGTH AND MICROSCOPIC APPEARANCE OF PAPER PREPARED FROM OLD CORRUGATED CONTAINER (OCC) PULP. BioResources 3 (2008) 383-402.

[82] C.J. Ridgway, and P.A. Gane, Size-selective absorption and adsorption in anionic pigmented porous coating structures: case study cationic starch polymer versus nanofibrillated cellulose. Cellulose 20 (2013) 933-951. https://doi.org/10.1007/s10570-013-9878-6

[83] F. Pereira, K.S. Sousa, G. Cavalcanti, M.G. Fonseca, A.G. de Souza, and A. Alves, Chitosan-montmorillonite biocomposite as an adsorbent for copper (II) cations from aqueous solutions. Int. J. Biol. Macromol. 61 (2013) 471-478. https://doi.org/10.1016/j.ijbiomac.2013.08.017

[84] Y. Sun, S. Zhou, S. Pan, S. Zhu, Y. Yu, and H. Zheng, Performance evaluation and optimization of flocculation process for removing heavy metal. Chem. Eng. J. 385 (2020) 123911. https://doi.org/10.1016/j.cej.2019.123911

[85] Y. Sun, S. Zhou, W. Sun, S. Zhu, and H. Zheng, Flocculation activity and evaluation of chitosan-based flocculant CMCTS-g-P(AM-CA) for heavy metal removal. Sep. Purif. Technol. 241 (2020) 116737. https://doi.org/10.1016/j.seppur.2020.116737

[86] P. Wu, J. Yi, L. Feng, X. Li, Y. Chen, Z. Liu, S. Tian, S. Li, S. Khan, and Y. Sun, Microwave assisted preparation and characterization of a chitosan based flocculant for the application and evaluation of sludge flocculation and dewatering. Int. J. Biol. Macromol. 155 (2020) 708-720. https://doi.org/10.1016/j.ijbiomac.2020.04.011

[87] Y. Sun, A. Chen, W. Sun, K.J. Shah, H. Zheng, and C. Zhu, Removal of Cu and Cr ions from aqueous solutions by a chitosan based flocculant. Desalin. Water Treat. 148 (2019) 259-269. https://doi.org/10.5004/dwt.2019.23953

[88] H. Wei, B.Q. Gao, J. Ren, A.M. Li, and H. Yang, Coagulation/flocculation in dewatering of sludge: A review. Water Res. 143 (2018) 608-631. https://doi.org/10.1016/j.watres.2018.07.029

[89] Y. Sun, A. Chen, S. Pan, W. Sun, C. Zhu, K.J. Shah, and H. Zheng, Novel chitosan-based flocculants for chromium and nickle removal in wastewater via integrated chelation and flocculation. J. Environ. Manage. 248 (2019) 109241. https://doi.org/10.1016/j.jenvman.2019.07.012

[90] L. Chen, Y. Sun, W. Sun, K.J. Shah, Y. Xu, and H. Zheng, Efficient cationic flocculant MHCS-g-P(AM-DAC) synthesized by UV-induced polymerization for algae removal. Sep. Purif. Technol. 210 (2019) 10-19. https://doi.org/10.1016/j.seppur.2018.07.090

[91] Z. You, C. Zhuang, Y. Sun, S. Zhang, and H. Zheng, Efficient Removal of TiO2 Nanoparticles by Enhanced Flocculation-Coagulation. Ind. Eng. Chem. Res. 58 (2019) 14528-14537. https://doi.org/10.1021/acs.iecr.9b01504

[92] S. Jia, Z. Yang, W. Yang, T. Zhang, S. Zhang, X. Yang, Y. Dong, J. Wu, and Y. Wang, Removal of Cu (II) and tetracycline using an aromatic rings-functionalized chitosan-based flocculant: enhanced interaction between the flocculant and the antibiotic. Chem. Eng. J. 283 (2016) 495-503. https://doi.org/10.1016/j.cej.2015.08.003

[93] H. Ge, H. Chen, and S. Huang, Microwave preparation and properties of O-crosslinked maleic acyl chitosan adsorbent for Pb2+ and Cu2+. J. Appl. Polym. Sci. 125 (2012) 2716-2723. https://doi.org/10.1002/app.36588

[94] M. Sugimoto, M. Morimoto, H. Sashiwa, H. Saimoto, and Y. Shigemasa, Preparation and characterization of water-soluble chitin and chitosan derivatives. Carbohyd. Polym. 36 (1998) 49-59. https://doi.org/10.1016/S0144-8617(97)00235-X

[95] X. Huang, H. Bu, G. Jiang, and M. Zeng, Cross-linked succinyl chitosan as an adsorbent for the removal of Methylene Blue from aqueous solution. Int. J. Biol. Macromol. 49 (2011) 643-651. https://doi.org/10.1016/j.ijbiomac.2011.06.023

[96] J. Ma, K. Fu, J. Shi, Y. Sun, X. Zhang, and L. Ding, Ultraviolet-assisted synthesis of polyacrylamide-grafted chitosan nanoparticles and flocculation performance. Carbohyd. Polym. 151 (2016) 565-575. https://doi.org/10.1016/j.carbpol.2016.06.002

[97] C. Fan, K. Li, Y. He, Y. Wang, X. Qian, and J. Jia, Evaluation of magnetic chitosan beads for adsorption of heavy metal ions. Sci. Total Environ. 627 (2018) 1396-1403. https://doi.org/10.1016/j.scitotenv.2018.02.033

[98] T. Lou, X. Wang, G. Song, and G. Cui, Synthesis and flocculation performance of a chitosan-acrylamide-fulvic acid ternary copolymer. Carbohyd. Polym. 170 (2017) 182-189. https://doi.org/10.1016/j.carbpol.2017.04.069

[99] L. Chen, H. Zhu, Y. Sun, P. Chiang, W. Sun, Y. Xu, H. Zheng, and K.J. Shah, Characterization and sludge dewatering performance evaluation of the photo-initiated cationic flocculant PDD. J. Taiwan Inst. Chem. E. 93 (2018) 253-262. https://doi.org/10.1016/j.jtice.2018.07.022

[100] W. Sun, H. Zhu, Y. Sun, L. Chen, Y. Xu, and H. Zheng, Enhancement of waste-activated sludge dewaterability using combined Fenton pre-oxidation and flocculation process. Desalin. Water Treat. 126 (2018) 314-323. https://doi.org/10.5004/dwt.2018.23076

[101] X. Lu, Y. Xu, W. Sun, Y. Sun, and H. Zheng, UV-initiated synthesis of a novel chitosan-based flocculant with high flocculation efficiency for algal removal. Sci. Total Environ. 609 (2017) 410-418. https://doi.org/10.1016/j.scitotenv.2017.07.192

[102] Y. Sun, C. Zhu, W. Sun, Y. Xu, X. Xiao, H. Zheng, H. Wu, and C. Liu, Plasma-initiated polymerization of chitosan-based CS-g-P(AM-DMDAAC) flocculant for the enhanced flocculation of low-algal-turbidity water. Carbohyd. Polym. 164 (2017) 222-232. https://doi.org/10.1016/j.carbpol.2017.02.010

[103] M. Tang, C. Zhu, Y. Sun, Y. Xu, H. Zheng, X. Xiao, W. Sun, H. Wu, and C. Liu, Preparation of polymeric aluminum ferric silicate for the pre-treatment of oily wastewater through response surface method. Desalin. Water Treat. 65 (2017) 284-293. https://doi.org/10.5004/dwt.2017.20314

[104] V. Haack, T. Heinze, G. Oelmeyer, and W.M. Kulicke, Starch derivatives of high degree of functionalization, 8. Synthesis and flocculation behavior of cationic starch polyelectrolytes. Macromol. Mater. Eng. 287 (2002) 495-502.

https://doi.org/10.1002/1439-2054(20020801)287:8<495::AID-MAME495>3.0.CO;2-K

[105] Z. Liu, M. Huang, A. Li, and H. Yang, Flocculation and antimicrobial properties of a cationized starch. Water Res. 119 (2017) 57-66. https://doi.org/10.1016/j.watres.2017.04.043

[106] M. Huang, Z. Liu, A. Li, and H. Yang, Dual functionality of a graft starch flocculant: Flocculation and antibacterial performance. J. Environ. Manage. 196 (2017) 63-71. https://doi.org/10.1016/j.jenvman.2017.02.078

[107] W. Sun, G. Ma, Y. Sun, Y. Liu, N. Song, Y. Xu, and H. Zheng, Effective treatment of high phosphorus pharmaceutical wastewater by chemical precipitation. The Canadian Journal of Chemical Engineering 95 (2017) 1585-1593. https://doi.org/10.1002/cjce.22799

[108] S. Mansouri, R. Khiari, F. Bettaieb, A.A. El-Gendy, and F. Mhenni, Synthesis and characterization of carboxymethyl cellulose from tunisian vine stem: study of water absorption and retention capacities. J. Polym. Environ. 23 (2015) 190-198. https://doi.org/10.1007/s10924-014-0691-6

[109] A. Shaabani, A. Rahmati, and Z. Badri, Sulfonated cellulose and starch: New biodegradable and renewable solid acid catalysts for efficient synthesis of quinolines. Catal. Commun. 9 (2008) 13-16. https://doi.org/10.1016/j.catcom.2007.05.021

[110] S. Hokkanen, E. Repo, T. Suopajärvi, H. Liimatainen, J. Niinimaa, and M. Sillanpää, Adsorption of Ni (II), Cu (II) and Cd (II) from aqueous solutions by amino modified nanostructured microfibrillated cellulose. Cellulose 21 (2014) 1471-1487. https://doi.org/10.1007/s10570-014-0240-4

[111] T. Suopajärvi, H. Liimatainen, M. Karjalainen, H. Upola, and J. Niinimäki, Lead adsorption with sulfonated wheat pulp nanocelluloses. Journal of Water Process Engineering 5 (2015) 136-142. https://doi.org/10.1016/j.jwpe.2014.06.003

[112] Y. Sun, M. Ren, C. Zhu, Y. Xu, H. Zheng, X. Xiao, H. Wu, T. Xia, and Z. You, UV-Initiated Graft Copolymerization of Cationic Chitosan-Based Flocculants for Treatment of Zinc Phosphate-Contaminated Wastewater. Ind. Eng. Chem. Res. 55 (2016) 10025-10035. https://doi.org/10.1021/acs.iecr.6b02855

Keyword Index

About the Editors

Dr. Vimal G. Gandhi
Associate Professor,
Department of Chemical Engineering,
Dharmsinh Desai University (DDU),
Nadiad-387 001 (Gujarat) India.
Email: vggandhi.ch@ddu.ac.in

Dr. Gandhi is, presently working as Associate Professor, with Department of Chemical Engineering, Dharmsinh Desai University since last 20 years. He received his Ph.D. in the area of Application of Nanotechnology in Environmental Engineering from Dharmsinh Desai University in 2011. He is serving as an Independent Director of BEIL Infrastructure Limited and Enviro Technology Limited (ETL) in Ankleshwar, Gujarat, India.

In the field of research and consultancy, he has guided several graduates and undergraduate students for their project work. He has more than 20 publications/presentation in international/national reputed journals and conferences to his credit. He edited one book on -"Photocatalytic nanomaterials for Environmental Applications" published by Material Research Forum, LLC- USA. His current interests are in the field of environmental engineering and synthesis of nanomaterials. He also organized various training programme for chemical industries in Gujarat including GNFC, PI Industries, Huntsman, Transpek Silox etc.

Dr. Kinjal J. Shah
Associate Professor,
College of Urban Construction,
Department of Municipal Engineering,
Nanjing Tech University (NTU),
Nanjing, China 218816.
M: +86 13072541186, Email: kjshah@njtech.edu.cn;
kinjalshah8@gmail.com

Dr. Kinjal J. Shah is an Associate Professor in the department of Municipal Engineering, Nanjing Tech University (NTU-China), he is also working as a visiting researcher at Carbon Cycle Research Center (CCRC), National Taiwan University (NTU-Taiwan). He completed his Ph.D. in the field of applied science from Graduate Institute of Applied Science, National Taiwan University of Science and Technology (NTUST), Taiwan in 2015. He has started his research carrier in 2009 from Shah Schulman Center for Surface Science and Nanotechnology (SSCSSNT), DDU, India. During his last 10 years of research carrier, he has published more than 30 SCI papers and 3 book chapters to International accredited journals. He has received leading Young Scientist award from Society of Polymer Science Japan in 2019, Japan. One of his research was nominated by ENI award, 2019, in category of "Advance Environmental Solutions". In addition, he has received many domestic and international awards from different societies and institutional bodies. He has been invited as lead speaker to many international accredited conferences.

His current interests are in the field of green chemistry and nano technology for sustainable environment. At present, his lab is developing technologies for advanced gas adsorption, and water purification technologies. He is serving as associate editorial board member in "Current Analytical Chemistry" journal, Bentham Science.